GLIMMER

... award-winning journalist and author who has written for *The New York Times Magazine*, *Wired*, *GQ*, *Reader's Digest*, *Los Angeles Times Magazine*, *Business 2.0*, and *New York Magazine*. His work appeared in the *2001 Best Business Stories of the Year*. He is the author of *Advertising Today*, *Hoopla*, and co-author of *Nextville* and *No Opportunity Wasted*, which appeared on *The Oprah Winfrey* show twice. He is also the creator and editor of *One*, an acclaimed national magazine focusing on advertising and design. His website is www.warrenberger.com.

Praise for *Glimmer*

Shortlisted for the CMI Management Book of the Year

'One of the best Innovation and Design books of the year'
BusinessWeek

'A different way of thinking about design – as a means of creative problem-solving' *Huffington Post*

'Berger's book has a relevance not only for people with an interest in design, but for anybody interested in rigorous, analytical thinking and problem-solving' *Irish Times*

'The premise of this book is that design is applicable to just about any challenge – and its principles are accessible to anyone' *CNN*

'These days, the buzz in business circles is all about "design". Gurus such as Am... ...n Berger have preached co... ...s nearly unlimited potentia... ...*and Mail*

GLIMMER how design can transform your business, your life, and maybe even the world

WARREN BERGER

BOOKS

Published by Random House Books 2011

2 4 6 8 10 9 7 5 3 1

First published in the United States by The Penguin Press,
an imprint of Penguin Group (USA) Inc. in 2009

First published in Great Britain in 2009 by
Random House Books
Random House, 20 Vauxhall Bridge Road,
London SW1V 2SA

www.randomhouse.co.uk

Addresses for companies within The Random House Group Limited can be found at:
www.randomhouse.co.uk/offices.htm

The Random House Group Limited Reg. No. 954009

A CIP catalogue record for this book
is available from the British Library

ISBN 9781847940056

The Random House Group Limited supports The Forest Stewardship
Council (FSC), the leading international forest certification organisation. All our
titles that are printed on Greenpeace approved FSC certified paper carry the FSC logo.
Our paper procurement policy can be found at www.randomhouse.co.uk/environment

Mixed Sources
Product group from well-managed
forests and other controlled sources
www.fsc.org Cert no. TT-COC-002139
© 1996 Forest Stewardship Council
FSC

Printed and bound in Great Britain by
CPI Bookmarque Ltd, Croydon, CR0 4TD

contents

The Briefing   *1*

SECTION I. UNIVERSAL

1. Ask stupid questŸns   *21*

   What is design? Who is Bruce Mau? And, by the way, does it have to be a lightbulb?

2. Jump fences   *45*

   How do designers connect, reinvent, and recombine? And what makes them think they can do all these things?

3. Make hope visible   *70*

   The importance of picturing possibilities and drawing conclusions

SECTION II. BUSINESS

4. Go deep   *99*

   How do we figure out what people need—before they know they need it?

5. Work the metaphor   *126*

   Realizing what a brand or business is really about—then bringing it to life through designed experiences

6. Design what you do   *155*

   Can the way a company behaves be designed?

SECTION III. SOCIAL

7. Face consequences   *183*

   Coming to terms with the responsibility to design well and recognizing what will happen if we don't

8. Embrace constraints   *211*

   Design that does "more with less" is needed more than ever in today's world

SECTION IV. PERSONAL

9. Design for emergence *239*

Applying the principles of transformation design to everyday life

10. Begin anywhere *267*

Why small actions are more important than big plans

The Glimmerati *293*

The Glimmer Glossary *301*

Resources *309*

Notes *315*

Acknowledgments *329*

Index *333*

THE BRIEFING

- You have a friend who uses a wheelchair. One day he tells you the three things about that wheelchair he hates: "I can't go over curbs, can't get up stairs, and can't look people in the eye when I'm talking to them." What would you do to improve your friend's wheelchair?

- Your grandmother keeps taking your grandfather's prescription pills (and vice versa). You tell them to check the labels on the bottles more carefully. They admit to you that they can barely even read the labels. How can you help them?

- You're designing a laptop computer for kids in the developing world. It cannot cost more than $150. But the parts alone cost more than that. What do you do?

- You'd like to get your teenage nephew to stop smoking. While you're at it, you'd like to get a million other teenagers to stop smoking, too. Telling them that smoking could kill them just makes it more cool and rebellious. So how do you make *nonsmoking* cool and rebellious, overnight?

- You're introducing a new automobile in the United States. Unfortunately, nobody's ever heard of this car. And it's precisely the opposite of what's been popular in cars lately. And it looks kind of funny. And there's no budget for television commercials. How do you make this car a bestseller?

- You find yourself in a village that desperately needs drinking water. There's a lake just a bike ride away, but the water's foul. How do you bring clean, drinkable water back to the village—using only your bike?

- You always wanted to do something creative and inspirational for a living. But somehow, you ended up operating a dog food company. How do you transform this into a higher calling?

- Your country has endured decades of misery brought on by two civil wars and an economy that keeps crashing. You need to turn all that around by creating a new spirit of optimism. What's your first step?

On the surface, these eight challenges might seem to be unrelated. They range from marketing issues to medical ones, from engineering to advertising, and from ambitious to seemingly impossible. But in each of these instances, the problem at hand was engaged by someone who looked at it in a fresh way and saw a possibility—just a glimmer, at first—of how things might be done differently. And in each case, that person went through a series of actions, guided by a common set of principles, which were all part of a larger established process. There is a word for that process: *design*.

When a New Hampshire man envisioned and then created a wheelchair that could stand and walk, he was designing. So, too, was the concerned granddaughter in New York who radically changed the look and shape of confusing medicine bottles, benefiting not just her grandparents but countless others. Principles of design led to the creation of a highly innovative (and irresistibly cute) laptop that rewrote the rules of low-cost computing, and those same principles helped turn the underdog Mini Cooper car into an automotive marketing phenomenon. Quite unexpectedly, design reversed a surge in teen smoking in Florida; in California it yielded a bike that could purify water as it was pedaled. Design enabled a dog food company to find new meaning and success by re-creating itself as an enterprise of, and for, dog lovers. And it was the secret weapon of a Toronto designer who organized a national movement that helped raise morale and productivity in the country of Guatemala.

When we think of design, we don't usually associate it with solving problems such as these. More often, it is equated with "style": fashionable clothing or handbags, distinctive typefaces, elegant Philippe Starck furniture or Michael Graves teakettles.

But design is really a way of looking at the world with an eye toward changing it. To do that, a designer must be able to see not just *what is*, but what *might be*. And seeing is only the beginning: Designers are also makers. They sketch and build, giving form to ideas. They take that faint glimmer of possibility and make it visible and real to others.

The process designers follow—which blends art and science and is fueled by human empathy—is arduous and at times heartbreaking. It is invariably filled with missteps, though each one tends to bring the designer a step closer to getting it right. And when that happens, the result can transform some aspect of the way we live. Suddenly, the act of listening to music, or peeling a potato, or accessing potable water is different, improved. In this way, progress happens by design.

The premise of this book is that design is applicable to just about any challenge—and its principles are accessible to anyone. If we can gain a better understanding of the ways designers think and work, it may enable us to do what designers tend to do so well: to recognize that glimmer of potential around us and within us, and to build on those nascent possibilities as we set out to design a better business or a better life.

DESIGN IS, according to Bruce Mau, "the human capacity to plan and produce desired outcomes." Mau, a renowned designer who agreed to open up his studio and its inner workings for this book, is at the forefront of a loose-knit movement embracing a new way of thinking about design. It includes individual designers and larger design companies like the leading-edge firm IDEO, as well as several prominent design schools, where new theories are being developed about "design thinking": what it is, how it works, what it can accomplish. But this "glimmer movement" also includes many from outside the design profession—including basement tinkerers, technologists, do-it-yourselfers, "crafties," social activists, environmentalists, video gamers, and business entrepreneurs. What links them is their belief that everything today is ripe for reinvention and "smart recombination." What makes them all designers is that they don't just think this, they act on it.

They are much better armed to do so than in the past. Aided by improved technology and connectivity, grassroots innovators can download instructions off the Web on how to make anything; they can share ideas and join forces with online collaborators halfway around the world; and when they've got a working prototype, they can stir interest by uploading it to YouTube. The democratization of design that began a quarter century ago with the introduction of Apple's Macintosh computer has moved to a whole new level in the era of interactive, social network media. It's true that much of the time these new "citizen designers" may simply be styling their own customized Nike shoes or Facebook pages. But as Bruce Mau discovered firsthand in his own research into the new design revolution, a growing number of people have set their sights higher. They're trying to design fresh solutions to old problems as they seek to improve the world around them.

Mau says the rise of people-powered design took him by surprise as he was doing research for his groundbreaking design exhibition called Massive Change. "I stumbled upon this movement that seemed to be bubbling up all around the world," he says. Everyday people were using design approaches and techniques to tackle thorny problems involving commerce, transportation, education, and housing (several of the examples cited at the start of this book were featured in Mau's original Massive Change exhibit). All of this suggested to Mau that design itself was undergoing a change—that it was moving beyond aesthetic concerns and the rarefied domain of design professionals and wading into the messy mainstream of everyday life.

For today's fix-the-world designers, there has been no shortage of challenges to tackle—indeed, the sense of urgency with regard to mounting environmental and economic concerns is a big part of what's fueling the movement. The problems themselves are testament to the importance of design—from levees not built to withstand the storm to financial systems that couldn't recover from a few critical shocks. As Mau points out, design is all around us, "but often it only becomes visible when it fails." And those now-obvious failures of earlier designs are prompting many to ask, *How can we reboot and rebuild—and do it better, more thoughtfully?*

THE NOTION that design can solve the world's problems is actually an old idea that has become new again. Going back over the past two centuries, a number of design movements—the Arts and Crafts Movement of the late nineteenth century, the modernism and futurism waves of the early and mid twentieth—have been fueled by an ambition to improve life for the larger population. The roots of design are entwined with utopianism and with dynamic figures such as the British designer William Morris, an early social-ist and a leader of the craft movement, or America's Buckminster Fuller, the dome-building dreamer who also was an environmentalist about eighty years before green became the new black.

By the 1980s, design had strayed pretty far from those utopian roots—unless, that is, one's idea of utopia is a world filled with pricey espresso machines. The design writer Phil Patton pegs the eighties as the time when designer-brand items, from jeans to high-end appliances, became the mark of good taste. Philippe Starck's Juicy Salif—a spidery, gorgeous, and expensive citrus juicer that seemed to be designed more for dramatic effect than for actual effectiveness—has been cited by Patton and others as the epitome of the "design for design's sake" mania that took hold. A decade later, Target stores upped the ante by "democratizing design"—which, in this particular use of the phrase, meant that more people could afford and have access to the designs of Starck, Graves, Karim Rashid, and other design luminaries, leading to what one observer called "the arms race of designer toilet brushes."

The growing appetite for stylishly made objects turned the Starcks and Rashids into the new rock stars, whose work could command huge sums (in 2006, a chaise lounge made by designer-of-the-moment Marc Newson sold for more than $900,000). But more recently, the notion of design for design's sake began to lose its luster—such that by 2008, Starck, in announcing his intention to shift focus to more useful things such as wind turbines, declared, "Everything that I designed is absolutely unnecessary."

It's little wonder Starck would begin to feel that way: The design realm of superstars and million-dollar chairs felt profoundly out of sync with what was happening in the world by the mid- to late 2000s. Rising environmental concerns had begun to stir the first rumblings of a backlash against all the nonessential "stuff" proliferated by designers, much of it disposable but not recyclable. As the economy contracted, those overdesigned and overpriced goods came to be seen as not just inaccessible but inappropriate. The world seemed to be crying out for a different kind of design: no less creative but more responsible and resourceful, with a greater emphasis on solving problems, embracing constraints, and doing more with less.

To some extent, this harkens back to earlier modernist design ideals, but with an important difference. While today's glimmer designers may think in terms of reinventing and somehow improving the world around them, they're more likely to approach the task humbly, and usually in an open, collaborative manner; there's less of the kind of hubris that might attempt to impose a grand, solitary vision upon everybody else. The working model developed by the design thinkers at IDEO or at Stanford University's influential "d.school" program calls for ongoing step-by-step evaluation and reconsideration throughout the design process, incorporating feedback from many people—including, perhaps most importantly, from those who will ultimately have to live with the results. Before they envision and sketch, and certainly as they proceed to build, today's designers do a good deal of watching and listening.

IT IS that aspect of design—the endeavor to first understand what is actually needed out there in today's world, before trying to satiate a need—that makes it relevant to business, particularly in the current difficult environment. Of course, design is nothing new to business, but the main application of it, until recently, has been as a means of trying to distinctively dress up product offerings that have exploded in number, with nearly two hundred thousand new ones introduced each year. In the quest to appeal to a more fickle and demanding public, companies have used design to tinker with the functionality and form of new products—a new color here, a new shape there, and a lot of new features everywhere (leading to the condition known as "featuritis"). But missing from the design of many of these products was a sense of purpose—which is not all that surprising, given that the companies behind the products were largely clueless about what the public actually *needed*.

There was a time when companies could survive and even prosper while making marginally useful, "me-too" products. The secret weapon was aggressive marketing and advertising. But with vocal consumers gaining more control of media and with a recession that has forced both companies and their customers to make every penny pay off, the air has leaked out of the hype balloon. At the same time, companies' ongoing attempts to downsize and to "reengineer" the old business processes in order to squeeze out more productivity have reached a point where there's nothing left to squeeze.

To grow now, companies must innovate and perform on every level, and that's where design comes into play. By employing design tools such as empathic research to uncover the quirks and foibles of how people go about their daily lives, design-driven companies are striving to close "the innovation gap"—the growing chasm faced by companies with the technical wherewithal to produce just about anything but with no idea of what to make. Beyond rethinking product offerings, companies can apply design to the ways they serve customers long after the sale, as well as the overall manner in which they conduct business—right down to the way phones are answered and trucks are routed. The whole experience can and should be designed, holistically.

According to Roger Martin, dean of the Toronto-based Rotman School of Management, "Design is becoming an ever more important engine of corporate profit." Indeed, design-driven companies have been shown to financially outperform others—in some cases, by as much as ten to one. Which is why Martin believes that in the increasingly tough business marketplace of today and tomorrow, "Businesspeople must begin to think like designers."

MARTIN IS right about that, but it's not just businesspeople who stand to benefit from adopting this way of thinking. At times, and perhaps especially in these times, we all need to reexamine assumptions about how and why we do things the way we do. Just as companies can rethink and transform their approaches to doing business, people in the social sector can and must find innovative ways to take on old challenges. Can we redesign the ways we provide a better life for the elderly, or encourage underprivileged children to learn, or shelter the homeless? The answer, as evidenced by the work of some of the designers featured in this book, is clear: Yes, we can.

On a personal level, good design is just as relevant—and we're not talking about coffee tables or handbags, but about the overall design of one's life: the choices made along the way, and the constant adjustments and refinements that ensue. For most of us, that overarching life design may be invisible, but it's there nonetheless. The question is whether that design is thoughtful or haphazard; whether it's expansive or restrictive; whether it's sturdy and long lasting, or temporary and fragile.

It can be difficult to step back and look at one's life with a fresh eye, but this is part of what design can teach us: how to view things sideways, how to reframe, rearrange, experiment, refine, and—maybe most important of all—how to ask "the stupid questions" that challenge assumptions about the way things have been done in the past. The design process is geared to breaking out of old patterns of thinking and behavior. That is not easy to do because of what designers refer to as "heuristic bias"—our natural inclination to think and act in familiar, repetitive ways, simply because the mind wants to follow a path that has already been cleared. Design thinking can help lead you off that well-worn trail and guide you in new directions.

TO LEARN to think like a designer, it makes sense to employ a tool that designers themselves use extensively: observational research. But trying to get an inside view of how design works can be challenging for a number of reasons. To begin with, the design world is splintered and fractious. The broad term *design* covers myriad disciplines, including graphic design, industrial design, architecture, fashion, environmental design, Web design, and more—and each of these disciplines has its own set of practices and principles.

At the same time, the practice of design has maintained a certain mystery about what it is and how it works. Even the corporate clients who hire designers often don't know quite what they're paying for (and, too, it has been said that designers' mothers have no clue what their sons and daughters do for a living). The graphic designer Alexander Isley says he has always felt that being a designer was "almost like being part of a secret guild." Another designer, Brian Collins, notes that his professional peers have a tendency to overcomplicate the discussion of what they do. "The design world is plagued by pseudoacademic jargon," Collins says, "with all these people talking about *mutable hierarchies of multivalent meaning*."

Bruce Mau has been grappling with the challenge of demystifying design for some time. Born and raised in a hardscrabble Canadian mining town, Mau, now fifty-one, burst onto the international graphic design scene in the 1980s, when his Toronto-based studio became known for its strikingly unorthodox use of type and images, featured primarily in the esoteric Zone books about culture and society. Over the years, Mau kept expanding his applications of design. He moved from the printed page to large public spaces—the Walt Disney Concert Hall in Los Angeles, the Seattle Public Library—where, working alongside the star architects Frank Gehry and Rem Koolhaas, Mau became a design star in his own right.

Then, a few years ago, Mau opted to widen his scope even more, as he began to view "the world" as his design project. In his work, his writings, and his exhibitions, he proposed that design principles could be applied to the thorniest global issues. When Mau debuted the project known as Massive Change in 2004—it began as a museum exhibit in Canada and then moved to America—he seemed to become the embodiment (and, at times, the lightning rod) for the new "design can do anything" philosophy.

In Massive Change, Mau showcased original designs from around the world that took on all manner of problems. There was the iBOT "walking wheelchair" designed by Mau's friend Dean Kamen; the low-cost laptop created by a group that included the noted designer Yves Behar; various wondrous devices that purified water or replaced lost body parts; new types of shelter, new ways of constructing cities—all of this was part of the show's purview. And the show's message was, in essence: *Yes, we've got large problems and challenges in the world, but there are answers, too. They're all around us, if you just look.*

Meanwhile, Mau also tried to take some of the fundamental design principles developed in his studio and go public with them. He took the first step when he wrote a document titled "An Incomplete Manifesto for Growth," which started as an article and then was given as a speech at an international design conference in the late 1990s. In the presentation, Mau laid out forty-three "laws" for achieving good and meaningful design. While he's a big believer in random experimentation and blue-sky creativity, Mau also believes that such creativity works best when guided by a proven design methodology. The manifesto quickly became a viral phenomenon on the Internet, and today, a decade later, its laws still are passed back and forth on the Web, mostly between young designers. Many of the ideas in the manifesto are counterintuitive. Mau advised designers to "ask stupid questions," to sometimes shamelessly imitate the work of others, and to always, at all times, "forget about good."

In his own work, Mau applies these principles to projects that have become increasingly complex and somewhat amorphous: helping the country of Guatemala figure out how to design a better future for itself; helping big American companies, such as Coca-Cola, design more sustainability into their products and processes; finding new ways for the television network

MTV to remain relevant to the youth culture in the age of YouTube. Mau has been enlisted by the president of Arizona State University, who wants to reinvent college education and is intrigued by Mau's radical ideas on the subject. He's been asked to redesign a Central American country, a major city in the Middle East, and a small town in Canada—tackling problems from traffic congestion to economic woes. (Perhaps the strangest proposal came not long ago from Colombia, whose government asked for Mau's help in an effort to rehabilitate former drug-dealing cartel members and welcome them back into society as functioning citizens. The project has since been shelved, but Mau was tickled by the prospect of taking on such a challenge—and fully aware of the potential hazards. Attempting to socialize gang members and drug dealers: How would a designer even begin to approach something like that? Mau's answer: "Very carefully.")

All of these diverse projects—from Fortune 500 companies going green to drug dealers coming clean—somehow find their way to the Bruce Mau Design studio, a bright and airy fourth-floor loft that overlooks Toronto's Chinatown district (Mau has also opened and is now based in a second studio, in a high-rise with panoramic views of Chicago). The workshop buzzes with activity as the staff designers—an ethnically diverse mix of people with backgrounds in art, filmmaking, engineering, architecture, and new media—work on a range of jobs at various stages of development, from scribbling on the walls to three-dimensional models and prototypes, with the heavyset, bushy-bearded Mau presiding over it all as the in-house guru.

At the studio, Mau is constantly on the move. When he does stop to sit and study a design problem, becoming deeply absorbed in it, he tends to tap his foot or jiggle his leg. He's all bustling energy that cannot be contained. Mau loves surprises and incongruities, and when he encounters them he lets out a startlingly playful laugh—high-pitched, almost like a child's. He also loves to explain things, and as he does he often makes quick drawings on whatever scrap of paper is handy. Like a lot of designers, Mau is part artist and part engineer. The artist in him is always sketching little visions of how things might be, and the engineer then has to figure out how to make them actually work.

———

THE WORKING principles developed over the years in Mau's studio serve as the starting point for *Glimmer*. But Mau is one of a handful of designers who are profiled in depth, with many others appearing in supporting roles—about fifty designers in all. A couple of the featured designers were part of Mau's original Massive Change exhibit: Dean Kamen and Yves Behar, each a renowned leader in the design field, though they come at it from different angles. Kamen, perhaps best known for having created the much-discussed Segway scooter, is a brilliant engineer and inventor who never thought of himself as a designer until Mau convinced him that he fit the definition. Mau believes that an engineer becomes a designer when he/she truly begins to empathize with human needs and desires, instead of just making things work mechanically. By this standard, a designer is an engineer with a soul. And Kamen, who has dedicated much of his work to helping people constrained by poverty or by their own physical limitations, clearly fits the bill.

Behar, meanwhile, was central to the successful U.S. launch of the Mini Cooper, which broke all the rules of automotive marketing by emphasizing design over expensive advertising (this and other examples in the book make the case that design *is* the new advertising—albeit more efficient and far more relevant to people's lives). Behar also has served as the lead designer on the One Laptop Per Child program, which produced the dazzling little green computer that promised to change the world and did not—but may yet live up to that hope when Behar unveils the even more radical third iteration of the device in 2012.

Cameron Sinclair, another featured designer, is a Londoner who's set up shop in America but can often be found drawing sketches in the dirt somewhere halfway around the world, usually in a place that has been ravaged by natural disaster. Sinclair's Architecture for Humanity group specializes in rebuilding these devastated areas with innovative low-cost design approaches—a testament to the ways in which designers can substitute imagination for resources, working with whatever is available. Sinclair's designers once devised a way to make a temporary clinic in Africa from plants growing out of the ground; it was elegant, cheap, efficient, and—best of all—when the shelter was no longer needed, it could be eaten!

The tales of how these designers achieve breakthroughs are not unlike mystery short stories: Something is amiss, no one can figure it out, and it's up to our protagonist to solve the conundrum. Jane Fulton Suri, who pioneered design's use of empathic research to try to understand why people behave the way they do, was a trained psychologist whose journey as a designer began when the British government enlisted her to try to figure out why so many people were suddenly cutting off their own toes with lawn mowers. Fulton Suri, as we'll see in Chapter 4, got to the bottom of it all, then proceeded to spend the next two decades solving similar problems as the head of design research at IDEO. Another designer in the book, Alex Bogusky, had to unravel a different kind of mystery: Why don't teenagers care that smoking could kill them? And if they don't care about dying, what *do* they care about? Bogusky, like Fulton Suri, relied on empathic research and other design tools to arrive at his "glimmer moment"—the point when a life-changing idea crystallizes in the mind.

THOUGH THE designers in this book are shown tackling all different kinds of challenges, they tend to go through the same steps and adhere to similar principles along the way. This book is organized around a list of ten of those principles. Several of them derive from laws in Mau's original manifesto, while others have been created fresh or adapted from the philosophies of other designers featured in the book.

The ten principles are divided into four separate categories: Universal, Business, Social, and Personal. The rationale is that while certain design principles apply to just about any creative problem-solving situation, others seem to pertain more directly to business, social, or personal concerns. But there is a lot of crossover between the categories; whether one is designing in business, in the social sector, or in life, the challenges can be surprisingly similar.

The Glimmer Principles

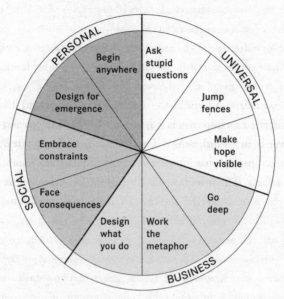

The **Universal** section of the book begins with a few basic design principles that might be used by anyone for any purpose—and that serve as the building blocks for all that follows. It starts with *Asking stupid questions*, which is really about learning to step back, look at things differently, and question conventional wisdom and accepted realities. That leads to *Jump fences*, or using abductive reasoning to envision fresh possibilities and forge new mental connections.

As those new possibilities begin to take shape in the mind, a designer will typically create a preliminary model—which can be as basic as a sketch on a napkin, or something far more elaborate and three-dimensional. It's a critical step on the journey to finding solutions. By visually expressing and sharing new ideas in rough form, the designer strives to *Make hope visible*, in the elegant words of featured designer Brian Collins.

The second section of the book is devoted to **Business**, where, as previously noted, design already has begun to turn the old MBA models upside down. In the business context, all of those principles from the preceding Universal section are still relevant, but another layer is added, on how to deal with the pressing challenge of differentiating products from one another in an overcrowded marketplace while also figuring out how to make brands and services come alive through meticulously choreographed consumer experiences.

If you've ever wondered how companies such as Apple or Target seem to have achieved such consistency in the way their brands look and perform, the answer is, in a word, *design*. They use it to seamlessly integrate every facet of their product development, marketing, store architecture, and customer relations. The principles *Go deep*, *Work the metaphor*, and *Design what you do* provide insights into how they do it—and these chapters also profile companies in the midst of transforming themselves, including Coca-Cola, Procter & Gamble, Pedigree dog food, and a small regional bank named Umpqua that has redesigned the process of banking so as to be unrecognizable. These companies understand that a recession is no time to retreat. It's actually a time to, in the words of one designer, "steer into the skid," by finding ways to efficiently innovate and refine a company's products, services, and its overall operations.

The third section of the book, **Social**, pertains to the use of design to address social issues and challenges. Today, designers are in the thick of efforts to clean up the environment, improve cities' infrastructures, respond to aging populations, smarten educational systems, and more. No one is suggesting that designers have all the answers to these complex and long-standing challenges—sometimes, in fact, designers can make things worse, particularly when they try to impose "solutions" without considering the consequences of those well-meaning actions. But when design is rooted in deep understanding of problems and local conditions, and when it carefully considers possible outcomes and takes care to *Face consequences* of its own making, it can offer real hope in difficult circumstances.

A hard reality of social design is that resources tend to be limited. We may want to fix the world's problems, but we usually don't want to spend much doing it. But the good news is that, interestingly, good design often doesn't cost any more than bad design. As resourceful and imaginative designers *Embrace constraints*, they find fascinating new ways of doing more with less. It can be a struggle to do so: Watching Yves Behar defy the perceived limitations of what's possible and fight to make his vision of that little green laptop into a reality, or seeing Cameron Sinclair toil to build something useful and even beautiful from the wreckage of disaster, should dispel any illusions that the act of designing social change is easy or bloodless. But it is possible.

The last section of the book, **Personal**, takes the discussion from the macro (the world at large) to the micro (your own life). With the mainstreaming of design, people are more inclined now to do their own interior decorating, create their own mix-and-match fashion sensibility, and even craft whole new identities for themselves in the virtual world. *I design, therefore I am* could be the new credo.

But the real personal design revolution is just getting started. Design used to be done *for* us—but increasingly it will be done *with* us or *by* us. People are already designing their independence from corporate life, their children's developmental experiences, their own lifelong education and growth, and, finally, their "golden years." The principles in this section, *Design for emergence* and *Begin anywhere*, focus on the ways that nonprofessional designers are using the process to solve mundane everyday problems such as How can I haul myself out of bed in the morning?, while also tackling much more profound questions: How do I express individuality? How do I balance living well and living responsibly? How do I design an environment that's conducive to creativity? How can I age comfortably? And lastly, is it possible to "design happiness"? (Short answer: Yes.)

WHETHER IT'S applied to business, social, or personal challenges, design thinking opens up new avenues of progress, suggesting fresh answers to old and difficult questions. It is about infinite possibilities. And perhaps more than anything else, it's about optimism. If there's one quality designers all seem to have in common (aside from a tendency to doodle on napkins), it is their optimism. Where many of us see troubles, they see opportunity—

because designers actually thrive on new problems, and the more difficult the better. This would seem to be a good attitude for our current times, with the continuous news cycle of gloom. It's too easy to become disheartened and even to become part of the problem. On this subject Mau likes to cite his friend, the author Stewart Brand, who has observed that when people believe they're in the midst of a crisis, they're more apt to behave selfishly; they circle the wagons.

On the other hand, Mau notes, if people believe they're living in a time of expanding possibilities, they'll want to be part of that growth. It's why Mau has a tendency to relentlessly point out that we're actually living in a time of unprecedented opportunity and virtually unlimited human capabilities. If that makes Mau seem overoptimistic, he has no problem with that. "A designer does not have the luxury of cynicism," he says. Mau believes designers must be willing to fall on their faces in pursuit of grand and ridiculous goals. "A cynical designer? That shouldn't exist," he says. "That's a joke."

More and more designers, along with a growing number of would-be designers, are starting to embrace the anything-is-possible message that Mau plastered on museum walls when he first launched his Massive Change initiative five years ago. Amid all the wondrous designs on display in that exhibit— startling yet modest human inventions that showed how people were using pure ingenuity and basic design skills to address daunting problems—Mau also hung a banner, which invoked the following challenge to all who entered: *What if we looked at the world as a design project—how might we begin to make it better?*

UNIVERSAL

I. ASK STUPID QUESTIONS

What is design? Who is Bruce Mau? And, by the way, does it have to be a lightbulb?

1.1 THE JOKE THAT EXPLAINS PROGRESS

As Dean Kamen tells the story, it started when he went out to get ice cream. "So I'm at the mall," says Kamen, who is a prominent New Englander now but who grew up in the New York City area and still speaks with the accent and clipped sentences of a New Yorker. "I'm on my way in from the parking lot, it's raining. I see a guy, in a wheelchair. He's not an old guy, he's young, fit looking—probably a vet, maybe had his leg blown off by a land mine, for all I know. But here he is, in this brand-new modern shopping center. And he can't get over the curb. He has to get help from a couple of other people, to lift his wheelchair over the curb."

Kamen raises an index finger to indicate: That's part one of the story.

"Few minutes later, I'm going by RadioShack, to get some batteries or whatever. I see the guy in the store, and now he's having trouble reaching something on the shelf. Then, as fate would have it, when I finally get around to going to the food court for my ice cream—there he is again! He's waiting to be served, but it's a high ice cream counter and he can't make eye contact. He couldn't do a basic transaction, not with any dignity anyway."

Now comes the point in the story when Kamen's design brain kicks into gear. "I'm looking at all this thinking, *What a pathetic lack of progress.* I mean, seriously—with all the incredible things we're doing with technology, what are we doing to improve this two-hundred-year-old wheelchair?

And what are we doing to restore this guy's dignity? Because that's what it's about—not just mobility, it's dignity. Can't we do better than this?"

And that was that. Because once Kamen raised that question—even if initially only within the confines of his own head—it begged an answer and set in motion a chain of developments. Kamen would spend the next several years trying to resolve dynamic stabilization issues in new ways, using solid-state gyroscopes, sensors, and microprocessors to simulate human balance in a package "small enough to sit under someone's butt," he says. "Because we knew once we could do that, we could stand a guy up on two points. And once we're balanced on two points, then we can deal with climbing the curb. And if we can climb the curb, then we can take it a little further and climb stairs."

Kamen eventually did manage to do all of that with his iBOT wheelchair: The seat raises its occupant to a standing position, as the wheels intrepidly roll up and over curbs or steps. The whole complex engineering and design project was triggered by a simple human observation and an emotional reaction to it.

"That's where our ideas always seem to come from," Kamen explains. "I think what happens is, we look at the same things, the same reality, as everyone else does. But we see it a little differently. Just because something is a reality today, we know that doesn't mean it has to be a reality tomorrow. So we're constantly looking at things and asking, Why? Or why not?"

AT THAT shopping center in New Hampshire, on a dreary, rainy evening, Kamen looked at a man in a wheelchair and saw the glimmer of possibility. But he was only able to see that by stepping back, reconsidering what he saw—and by questioning the way things are and might be.

The questions Kamen raised at the time—*Why shouldn't someone in a wheelchair be able to stand and make eye contact with others? Why can't he climb curbs, or even stairs?*—fit the definition of what Bruce Mau calls "stupid questions": the kind that challenge assumptions in such a fundamental way they can make the questioner seem naïve. Had Kamen asked these questions at, say, a business meeting within a company that makes wheelchairs, they very well might have elicited discreet eye-rolling and restlessness, along with a feeling that the meeting's forward momentum had ground to a halt.

But in actuality, the opposite is true. The act of questioning basic assumptions can be the first step toward reinvention and meaningful change. And it is often design's starting point.

Designers are so known for questioning that there is a joke acknowledging this tendency:

How many designers does it take to change a lightbulb?

Answer: *Does it have to be a lightbulb?*

Joking aside, when designers ask whether "it has to be a lightbulb," what they are doing is reframing a familiar problem or challenge in an unconventional way. *Framing* is a favorite term among designers, used with various meanings, but it generally refers to the way a problem or challenge is defined and laid out by a designer who intends to try to solve it. And often, the way a problem is framed will determine the solution. The problem of needing to figure out how to change that lightbulb may be reframed as a need to bring more light into the room without constantly *having to* change the bulb. This, in turn, may lead to putting a window in the roof to let the sun shine in.

The inclination to ask stupid questions and to frame problems in new ways (the two practices tend to go hand in hand) is a big part of what makes a good designer good, but it can also be useful behavior for just about anyone. To be able to step back, look at what surrounds you with a fresh eye, and question what is usually taken for granted is how people can change their lives, how societies or governments can retackle old problems, and how companies can regain focus or completely reinvent themselves.

Mau says that, as the economy soured in late 2008, he found he was getting an increasing number of phone calls from businesses who were discovering that in this new, difficult environment, the old formulas and models that these companies had lived by for years were no longer working. Some of them undoubtedly were hoping Mau might be able to hand them a new formula, a bit of design magic. But what these companies really must do, Mau says, is the hard work of reconsidering, which often requires that they ask themselves some very stupid (but by no means easy) questions as they try to reframe their very purpose: *Why do we make the things we're making? Does anybody still need this stuff? What if we were to radically change this thing we make? Or make something else instead? Maybe we need to stop making "things" altogether and start providing something more—a service, an experience?*

The questions are so fundamental that a lot of companies haven't stopped to consider them in a long time, if ever. Likewise, social services providers are using old models that are not holding up well in these times, but to adapt to a new reality, basic questions must be asked: *Never mind what we're used to providing; what do senior citizens really need these days? What makes a poor child want to learn? What do homeless people do all day?* And on a personal level, asking stupid questions is every bit as relevant: *Where should I really be living? How can I get more done? What makes me happy?*

DESIGNERS KNOW that asking fundamental questions is not easy to do, but on the other hand, you don't have to be an expert, either. In fact, generally speaking, experts are the ones least capable of asking stupid questions, because they know too much (or so they think). Designers, on the other hand, are often in the role of "anti-experts"—they tend to come at challenges from the perspective of the outsider. In business, designers are often brought in from outside to solve a problem, but even if the designer works in-house at a company, he/she is usually expected to take an "outside" point of view—one that is more in line with the end user or customer than with the company's executives.

Designers like Mau or Paula Scher from the design firm Pentagram are used to taking on assignments that range from working for a bank one week to a hospital the next. As they jump around, they're not expected to have deep expertise in each particular industry, and in fact their lack of inside knowledge can be a great asset. "When I'm totally unqualified for a job, that's when I do my best work," explains Paula Scher, a renowned graphic designer who has worked with public theaters, children's museums, and banks (she once sketched an umbrella logo on a napkin for Citibank and, in a matter of minutes, changed the identity of one of world's most powerful financial institutions—though, alas, Scher's umbrella couldn't keep Citi from getting soaked in the financial crisis). Scher says, "If you're trying to find a new way to think about something that makes it better, it can actually hurt you to have too much experience in that particular milieu—because you understand the expectations too well. And that can cause you to limit and edit your possibilities, based on what you already know 'doesn't work.' "

But if you're inexperienced in a given area—or, to use Scher's words, "if you're a complete neophyte, a moron"—you'll tend to ask questions that elicit more profound responses. "You'll ask what would seem to be the obvious, except nobody's seriously thought about it," she says. "From ignorance, you can come up with something that is so out of left field that it has been ignored or was never considered a possibility."

Questioning and framing require that one try to observe situations or scenarios in an open, unbiased manner. Mau sees this as akin to adopting a child's view of reality, in which everything is noticed as if for the first time and it's all subject to inquiry and investigation.

It makes sense, then, that the specific question that is often most useful in this approach is the one favored by inquisitive kids everywhere: *Why?*

It's such a good all-purpose query that many designers have adopted it as a kind of mantra used throughout various stages of their work. The design-driven medical equipment company Modo has been known to use cue cards, placed throughout the company, to continually remind company employees—designers and nondesigners alike—to ask "why" at every stage of conducting business. And the design firm IDEO has established a methodology practice known as the Five Whys. When the company is trying to arrive at new insights on a particular issue, its design researchers ask *Why?* over and over, in response to every answer they get.

Asking questions, especially fundamental ones, "forces you to be honest about what you don't know," says Clement Mok, one of the early creative directors at Apple. It can also put you in an uncomfortable position. The designer George Lois relates that he has found himself in the following

Example of IDEO's Five Whys Methodology

situation countless times: "You're in a conference room and everybody is nodding their heads, and for some reason you're the only one who raises his hand and says, 'Wait a minute, this thing you want to do doesn't make any sense. Why the hell are you doing it this way?' It takes some guts to be the one to raise your hand. It's easier to nod."

As Mau points out, "The fear for so many people is that, in asking these kinds of questions, they will seem naïve. But naïve is a valuable commodity in this context. Naïve is what allows you to try to do what the experts say can't be done."

The recent history of design innovation backs up Mau's assertion. So many stories of design breakthroughs, including the ones featured in this book, have begun with someone asking (much as Kamen did), *Why does it have to be that way?*

Two decades ago, when Sam Farber observed that his wife was having trouble peeling carrots because of her arthritic hands, Farber wondered why no one had designed a peeler with a handle that was easier to grip. Among the established makers of potato peelers and other low-cost kitchen items, this would be perceived as a "stupid question," and not in Mau's positive sense. After all, who could afford to cater to a small segment of customers with arthritis? And why would any customer pay a premium for an item no one cared about to begin with? Why bother to redesign something as mundane as a potato peeler?

Those last few questions, it turned out, were not stupid—they were just plain dumb. The new peeler introduced by Farber, with its thick, contoured, and tactile handle, appealed to a wide audience and propelled his new company, OXO Good Grips, to the top of the industry. Anyone who has used OXO products can probably guess that considerable design effort goes into making them feel so right you actually want to pick them up and do chores. Design principles involving ergonomics and usability inform every ridged surface or curved handle. But the real starting point for the peeler and so many of the company's products can be traced to someone asking why a common device couldn't be made a little bit better or why an everyday task couldn't be made easier.

Farber's experience also illustrates that these basic questions can be inspired simply by looking at what's going on around you, as was the case with Kamen and the wheelchair, or with Deborah Adler, who observed that her grandparents were having trouble using their medicine bottles and came up with an ingenious redesign.

Invariably, these innovations—whether achieved by seasoned professional designers or amateurs—weren't born in research labs, as one might expect. They started with questions first formed in kitchens, bathrooms, ice cream parlors, and other places where someone observed a human need and wondered why it wasn't being met. Once the question is raised, then begins the journey of envisioning, building, and refining an answer—the process of design.

That process will be examined and explored through many of the stories in the book, but it may help, at the outset, to have some basic context on design and designers. And that requires asking, up front, a few stupid questions about design itself, such as: *How does it affect us? How are designers different from the rest of us?* And the stupidest question of all: *What is design?*

1.2 A WORD THAT NEEDS ITS OWN DICTIONARY

When asked to define design, some of its practitioners have a hard time explaining what they do every day of their lives. No wonder their mothers are bewildered, to say nothing of designers' children: "All my kids know about what I do," says designer Alex Isley, "is that daddy plays with crayons all day."

The problem is not a lack of good working definitions, but rather an overabundance of them. For example, when designer Milton Glaser was asked for one, he offered up three (all good ones, but still—couldn't he have picked just one?). The facing page, designed by IA Collaborative, can be thought of as a top-twenty list of definitions, but the chart easily could have been expanded to a hundred or more.

That there would be so many definitions is understandable when you consider that the activity of designing takes so many forms and spans so many disciplines: fashion, graphic design, industrial or product design, Web design, interior design. The term has been picked up by landscape gardeners, cosmetic dentists, even life coaches (now called lifestyle designers).

DESIGN IS:

AN ASPIRATION TO CREATE

OUR PASSION TO HELP HUMANKIND

A STRATEGY TO EFFECT CHANGE

THE DESIRE TO IMPACT THE WORLD

OBSERVATION **figure 4**

"A means by which you see yourself and a means by which you express yourself to others." *Eliot Noyes*

"The art of making something better, beautifully." *Joe Duffy*

"Art that people use." *Ellen Lupton*

"The visual communication of (hopefully) a Big Idea." *George Lois*

OBSERVATION **figure 3**

"How you treat your customers." *Yves Behar*

"The act of giving form to an idea with an intended goal: to inspire, to delight, to change perception or behavior." *Clement Mok*

"The expression of an idea, process or system for the betterment of client interests and human locomotion not excluding the recent trend of lower x-height among Dutch typographers." *Stefan Sagmeister*

OBSERVATION **figure 2**

"A plan to make something, for a specific purpose, with a specific audience or user in mind." *Michael Bierut*

"The human capacity to plan and produce desired outcomes." *Bruce Mau*

"A plan for action." *Charles Eames*

"The introduction of intention into human affairs." *Milton Glaser*

"Moving from an existing condition to a preferred one." *Milton Glaser*

"The conscious and intuitive effort to impose meaningful order." *Victor Papanek*

"The art of planning." *Paula Scher*

"Above all, discipline." *Massimo Vignelli*

OBSERVATION **figure 1**

"Hope made visible." *Brian Collins*

"Anything that God didn't make." *Alexander Isley*

"The soul of a man-made creation." *Steve Jobs*

Then, too, there's the nonprofessional usage: If you rearrange your garage in a somewhat creative way, you're designing on some level. "As a society, we actually use the word more intelligently in the colloquial sense," Mau notes—which is to say, if you ask professionals what design is, they can get tangled up in all those overlapping disciplines, but a nondesigner usually gets that it's about figuring out something in advance and making a thoughtful plan to do it.

In a word, a design is a *plan*, but Paula Scher takes that definition to a loftier level, declaring that design is "the art of planning." One of Scher's creative partners at the Pentagram design firm, Michael Bierut, adds some specificity, defining design as "a plan to make something, for a specific purpose, with a specific audience or user in mind." Of the many definitions, perhaps the most specific one was offered by the quirky design star Stefan Sagmeister, who decreed that design is "the expression of an idea, process, or system for the betterment of client interests and human locomotion, not excluding the recent trend of lower x-height among Dutch typographers."

The definitions range pretty far, but the common threads seem to involve planning, purpose, and intent. And this brings up an important point about the distinction between art and design. Design can certainly be artful, but there's a critical aspect of it that often distinguishes it from pure art. Even the most beautifully designed objects are also usually meant to be functional; hence design becomes, in the words of writer/designer Ellen Lupton, "art that people use."

The sculptor Donald Judd gets to the hub of the difference with this line: "Design has to work. Art does not." (The painter David Hockney countered by pointing out that, conversely, "Art has to move you and design does not, unless it's a good design for a bus.")

There is a point when design fully crosses into the realm of art, becoming almost entirely about style and personal expression, with little emphasis on utility. At that point, all bets are off. In the rarefied world of "art design" (or is it "design art"?) most of the sturdy principles of design—all of those concerns having to do with human need and human nature, rooted in meticulous planning and refinement, all geared to getting something right or making it better—go out the window, along with lots of dollar bills. A chair ordinarily is designed to keep someone's backside from crashing to the floor, but if it's a chair sold to collectors at the design fair at the Venice Biennale, chances are no one will sit in it—so those structural principles no longer matter. Obviously, art design can be beautiful and expressive and delightful, which is a purpose in itself; but for the most part, it is separate from the kind of design that is engaged in the struggle to solve human problems—which is the kind of design that is the focus of this book.

THERE'S A disconnect between the way serious designers define design and the way the popular culture defines it. Notwithstanding all of those definitions referring to purpose and progress, when you examine the ways design is usually discussed in the media, the emphasis is overwhelmingly on style. Just check the television listings and the magazine racks. In terms of the former, there has been an explosion in recent years in the number of TV shows with "design" as a theme, but most of them feature fashion designers and interior decorators.

A few years back, the journalist Virginia Postrel wrote an interesting and persuasive book, *The Substance of Style*, which argued in favor of style and showed the importance of aesthetics in helping people enjoy their lives. Postrel's point: Style isn't merely superficial—it matters, deeply. That's true enough, but the trouble is that it often matters too much, particularly to the people making design decisions. Many designers have spent the better part of their careers trying to convince corporate clients, and the world at large, that design is not only about appearances.

Mau sometimes refers to the "tyranny of the visual" when discussing popular perceptions of design. He doesn't blame anyone in particular for this; he believes the emphasis on style evolved as a natural response to designed objects. "It's understandable, given the history of design as the shaping of found material," he says. Through the years, artisans and creators took hold of materials at hand, such as wood, stone, and clay, "and by shaping that matter we generated utility and delight," Mau says. "And the form, the shape of things, was the principal method by which design produced its wonders."

In the business world, the word *design* has been almost synonymous with *style*. Until recently, designers were tasked with making products look better and creating eye-catching packaging and communications. All well and good—those are certainly important functions. But the problem, at least in the minds of some designers, was that "style" became a kind of ghetto for them. Being associated with appearances made it difficult for designers to exert influence when key business decisions were made about what kinds of products should be produced and how they should work. That is beginning to change: "Clients used to come to us and say, 'Can you make this thing a little nicer or better?'" says Davin Stowell, founder of the firm Smart Design. "Now they're coming to us and asking, 'What should we be making?'"

The reason cosmetic design was emphasized in business was that it helped move merchandise—offering a way to take mass-produced goods that were similar to one another and spiff them up so they'd stand out. It could make "this year's model" seem new and improved, even if it really wasn't. And if a product was designed in a way that made it look better, you might enjoy it more and might even come to believe that it was improved—all because of its appearance. As has been noted by Postrel, as well as by the design guru Donald Norman and others, the aesthetic aspects of design can go beyond superficiality in terms of impact on people. Norman, in particular, has theorized that style is interlinked with functionality and that it all comes together to influence us on three distinct levels.

1.3 A) "I LUST AFTER IT" B) "IT IS SMART BUT I'M SMARTER" C) "IT COMPLETES ME"

Norman is an astute observer of design who teaches at Northwestern University and heads up the Nielsen Norman design consulting firm, which has worked with Apple and other leading design-driven companies. A cognitive scientist by training, he wandered into design after first wading into a disaster. Norman had been called upon in the aftermath of the Three Mile Island nuclear accident to study the situation from the perspective of a brain scientist. It was hoped that he could shed light on why the power plant's workers seemingly had failed to react adequately to the crisis as it was unfolding. But Norman found that the fault lay not in the neural paths and synapses of the workers—it was in the poorly designed buttons and graphics of the plant's control systems.

From then on, Norman became hooked on the science of design. He has been studying, ever since, the ways that people react to designed objects and environments. His three-level theory holds that design appeals to us on:

- The visceral level, wherein we respond mostly to appearances.

- The behavioral level, where usability comes more into play (Do we feel at ease operating something? Can we "master" it? Does it seem useful?).

- The reflective level, which is tied up with issues of identity, self-worth, and intellectual appeal—as in: Do I want to be associated with this product? Do I feel good about owning it? Can I tell stories about it and impress others?

3 Levels of Design Appeal

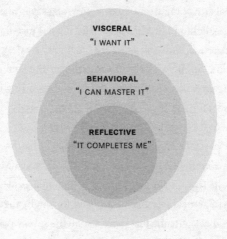

VISCERAL
"I WANT IT"

BEHAVIORAL
"I CAN MASTER IT"

REFLECTIVE
"IT COMPLETES ME"

Which of the three levels is the most powerful? "If any one of those is going to trump the others, it's the reflective level," Norman says. "It's all about the image we project. People will go out of their way to buy things that fulfill and enhance that image."

When a design connects on multiple levels it can become a phenomenon, like Apple's iPod or iPhone. In the case of the iPod, Norman notes that the product is actually far from perfect from a functional standpoint. "People have trouble turning it on and off, and they get frustrated because they can't replace the battery," Norman points out. But they forgive all that because on the visceral level the product is sleekly attractive, while on a functional level it allows people to easily manage their music, and on the reflective level we all want to be part of the "different and smarter" club that Apple represents.

To boil down the three levels a bit, the ideal response to a designed product might go something like this: 1) "Wow! How cool looking is that?" 2) "Hey, this is easy. I can do this. And I can picture myself using this every day." 3) "I can't wait to show this off to my friends."

As Norman says, from a psychological standpoint number three may be the most powerful level. But in terms of opening the door and getting things started, visceral appeal is critical. And this is where aesthetics plays its biggest role. Robert Wong, who served as the design director at Starbucks as the chain was becoming a phenomenon, compares the visceral appeal of design to sexual attraction. Beautiful design, Wong says, "makes the medial prefrontal cortex light up, just like sex." And it's not just a matter of visual impact: Sound, touch, and even smell can be utilized by designers to get the hormones racing. Starbucks learned early on that the sound and smell of the beans being ground actually made the cash register ring more.

SOME OF our visceral responses to design seem to be hardwired. According to Moshe Bar, a cognitive neuroscientist at the Harvard Medical School, research suggests that we simply respond better, on a gut level, to smooth, curved, and symmetrical designs as opposed to rough, angular, jagged, or uneven shapes. Bar contends that there's something very primal in all this:

When we're around smoothly designed, aesthetically pleasing objects, we feel less threatened. Conversely, rough edges and sharp contouring tend to increase activity in the amygdala area of the brain, which responds to anything perceived as a possible cause for concern. Cluttered, messy designs also can be unsettling. "The brain doesn't like uncertainty," Bar says, "and if there's too much going on in a product's design, we may associate that confusion with possible danger."

It's logical that style and aesthetics would be important at the visceral level, but what's more surprising is this: *If something is beautiful, it may be easier to use.* That, at least, is what has been suggested by studies (cited in Norman's book *Emotional Design*) showing that people do a better job of figuring out and using products that look good—apparently because an aesthetically pleasing object causes you to relax more as you use it and also to be more patient and more creative as you try to figure it out. If something looks like it's badly or cheaply designed, we're more apt to throw it against the wall than to work through frustrations.

That's a good lesson for designers engaged in all kinds of endeavors, including would-be world-changers: Even if you're trying to design something profound and noble, it still benefits from being sexy. When Yves Behar designed the XO laptop for children in the developing world, he slaved over the cute and colorful appearance of it. Behar wanted kids who were absolutely clueless about computers to covet these devices before even knowing how to switch them on.

Once they do turn it on, that's when things really start to get interesting from a design standpoint. Because at this next level, the behavioral one, the designer must somehow come to terms with other people's needs and capabilities—and then design something that matches up to those needs and capabilities. The result, at times, can seem to be the work of someone with ESP, though in fact it involves at least as much science and sweat as intuition. Designers rely on precise principles such as mapping to guide the user of a product or service in the right directions, and they build in "forgiveness" (a wonderful name for a design principle) to help us avoid mistakes and recover easily from them. So many of these neat tricks are completely invisible to the user. But that just adds to the appeal because it allows us to believe, as we use well-designed products, that we're the ones who've mastered the complexities and figured it all out. For a few moments at least, we're in control and the world makes sense.

1.4 GRADUATING FROM OBJECTS TO OBJECTIVES

In marveling at how designers, at their best, can make things work so well and so easily, one can't help asking a provocative (and, yes, slightly stupid) question: Why can't everything be designed as well as an iPod? And this is where Bruce Mau reenters the picture, because he believes that, in fact, everything can.

Some time ago, Mau pretty much stopped designing objects (books had been his object of choice for years) in part because he began to feel design offered greater possibilities when it came to changing or "transforming" what already exists, as opposed to making shiny new geegaws for the shelves of design stores.

In the design world, there has always been "a constant tension between inventing and improving," says the designer Valerie Casey from the firm IDEO. Invention has tended to bring more money and glory to designers—indeed, the pressure's always on to create the next hot-selling product, and that won't change. But at the same time, there's a growing recognition of the need for smart redesign, or designed improvement— whether in existing products (such as by making them more sustainable), in companies and how they operate, in social services, or, generally, in the way the world works.

What's interesting is that a lot of the same principles that apply to designing "stuff" also apply to this other category of design—the design of improvements or solutions. In both cases, there is the same need to start by questioning conventional wisdom and accepted practices; the same imperative to uncover what it is that people most need (even if they don't know they need it) and also what they're likely to respond to; the same practices of creating visual models or physical prototypes of new ideas, and then refining those based on feedback.

But this new strain of design—which sometimes goes by the currently popular term "transformation design" or by Mau's own branded version of it, "Massive Change"—tends to be grander in scale and more complex than basic product design. Perhaps the biggest difference is that it calls upon designers to move beyond creating "things" and to begin orchestrating "experiences."

One could argue that designers have always been in the business of creating experiences. The very best "object" designers, from Charles Eames designing furniture to Raymond Loewy making household appliances, understood that they were, on some level, designing the experience of living with an object. Certainly, they sought to empathize with the user, to understand the context in which the product might be used, and even to direct the user's experience somewhat by building in certain intuitive features.

But there is a world of difference between designing a comfortable chair and, for example, redesigning the way companies can make their practices more sustainable, or the way hospital stays can be improved, or the way cities can transport people more efficiently. While the design of the chair results in something tangible and sturdy (unless it's one of those fancy Biennale chairs), the efforts of transformation designers can yield far more amorphous results. But the payoff, if and when it works, is potentially huge.

As Mau and other transformation designers try to push the limits of what design can do and what it encompasses, that basic question *What is design?* gets harder to answer—as does the question *What does a designer do?* Not everyone is happy about this. As the transformation design trend first started to take hold in the United Kingdom a few years ago, Mike Dempsey at the Royal Designers for Industry trade group complained to a British newspaper that the term *designer* was being stretched and "abused." Dempsey pleaded, "Can we please have our name back?" In the United States, Pentagram's Michael Bierut has lamented the trend, wherein, he wrote, "Design is presented as a kind of transformative cure-all, with the designer as scold to tell you What You're Doing Wrong."

The term *scold* doesn't quite fit Mau, and he certainly doesn't claim to have all the answers—he has a lot more questions than answers. Still, it is fair to wonder just what kind of person feels qualified to take on a redesign of the world. Just who are these people to tell us what we truly want and how we really ought to be doing things?

1.5 INVASION OF THE T-SHAPED PEOPLE

The designer Milton Glaser tells a story about his childhood that provides an insight into how designers tend to see the world around them. The young Glaser fell ill for an extended period of time and was bedridden. "Every morning," he recounts, "my mother would bring in a wooden board and a big lump of clay. And I would spend the day making complete villages—I would have this vista in front of me. Then, at the end of the day, I would take a mallet and pound it all back into a big lump of clay—excited with the prospect that I'd have to do it all over the next day. I think that was essential to my survival, because it gave me something to look forward to: Each day I got to recreate the world."

It's not uncommon to find designers who grew up constructing their own worlds, sometimes as young solo inventors tinkering in the basement. For instance, as a teenager Yves Behar wanted to combine skiing and surfing in one piece of equipment, so he created a board that could windsurf on frozen lakes. The maverick British designer Thomas Heatherwick was, by age six, making sketches of a toboggan with its own built-in suspension system because, he explained years later, "I didn't see why my bum had to hurt." Dean Kamen designed audio/video light shows in his parents' basement (he was so good at it that while still a teenager he was hired by New York museums to design their light shows).

But designing professionally is usually not a solitary occupation—the designer, unlike a solo artist, must meet with clients and collaborate with coproducers of projects—and so even those designers who start out as somewhat introverted young creators usually turn outward as adults. It makes for a dichotomous breed: They are the cool nerds, the plugged-in outsiders.

The best designers seem to be interested in everything, to the extent that it can make them seem unfocused. John Maeda, president of the Rhode Island School of Design and a renowned designer himself, says that through college and his early career, "I was doing so many different things"—Maeda was interested in graphic design, computer science, physics—"my teachers would say, 'John, just do one thing.' It's the curse of the curious. It's seen as being wrong somehow."

But it's not wrong for designers, because, in fact, they need to be generalists given the scope and range of the challenges that are brought to them. As Pentagram's Bierut puts it, "Not everything is about design, but design is about everything."

Even with all their hopping around, most designers are trained in a particular design skill set—say, graphic design or industrial design. According to IDEO's chief executive Tim Brown, the best designers are what he calls "T-shaped people." This means they start out with deep interest and expertise in one skill—that's the vertical base of the T—and then, as they blossom as designers, they branch out into many different areas of knowledge. Throughout their careers, Brown says, both parts of the T should keep growing together. "So this means that, ideally, people might come to IDEO as, say, an 8-point T and gradually become a 64-point T."

T **T**

The broadening of the top of the T becomes increasingly important in the new world of transformation design, according to Brown. "To respond to the complexity of design problems today, what's required is not one person working alone in a studio—it's more likely to be interdisciplinary teams, and those people need to be able to work effectively together," he says. "What we've found is, if someone has an enthusiasm or curiosity about many different disciplines, then they can be more flexible, more empathetic, and more engaged with the world."

As T-shaped people venture wide and dip into many different milieus, the empathy cited by Brown enables them to view things from the perspective of others. But at the same time, they're always seeing the world through the meticulous eye of a designer. They can't help it: Everywhere they look designers see design, in all its glory and, more often, with all its flaws. Once, when the designer/architect Michael Graves found himself laying on a gurney in a hospital, the words that came out of his mouth were: "I don't want to die here—it's too ugly."

1.6 THE BOY WHO JUMPED OVER THE FENCE

In many ways, Bruce Mau fits to a T the prototypical design personality that Brown has described. Which is not to say that there is anything typical about Mau's background. As he points out, "I am probably one of the few designers you'll meet who can put a pig in the freezer for you if you need it."

Mau grew up on a small family farm just outside the nickel-mining town of Sudbury, several hours north of Toronto, in one of Ontario's colder latitudes. While his stepfather toiled in the mines, Mau and four sisters worked the farm, which had no running water in the winter. Life in Sudbury was "cold, harsh, nasty," Mau says.

From as early as he could recall, Mau wasn't interested in the same things as everyone else. "I didn't like to shoot things," he says, and when it came to his athletic prowess, "the hockey gene skipped a generation." He did like photography, art, and science.

Away from the farm, Mau's place of refuge was a student science lab at the local high school, where he buried himself in projects such as making his own radio. He figured he'd be a scientist. "But one day," he says, "as I was playing with a homemade radio I thought, 'You know, I just don't want to do this forever.' So I put the radio down and went to the guidance office

BRUCE MAU:

Imagine Sudbury

A few years ago, I was invited by the local chamber of commerce to deliver a lecture in my hometown of Sudbury, a mining town in northern Canada. As a child I lived on a farm in the northern forest outside the city, and every day as I took the bus to school in the center of the city, I passed through a thirty-mile dead zone where barely a blade of grass could survive. The pollution from extracting and processing nickel was so extreme that a landscape of extraordinary beauty and richness had been rendered bleak, mostly just sand and rock. It was rumored that when NASA wanted to train astronauts to drive on the moon, they were brought to Sudbury.

But at my lecture, I started off by showing a very different image: paintings by the Group of Seven, the famous Canadian artists who traveled to the Sudbury region from Toronto to paint the spectacular beauty of the northern wilderness. I presented these images of a place with bountiful lakes and rivers and endless forest, and said: "So, this is what we have to work with—and this is what we do with it. . . ."

Now the image on the screen was of four typical Sudbury street corners. Cold and gray, harsh and desolate. These places seemed devoid of design, culture, or life—nothing but parking lots and median strips.

There was an audible gasp. They were embarrassed.

I asked the assembled leaders of Sudbury: What message did they imagine young people were receiving when they encountered this environment? To me, it was: "We are not willing to make this place somewhere you want to be. If you are talented, intelligent, and mobile, you might as well get out of here. Find somewhere beautiful."

I proposed a four-part strategy to radically reimagine this place.

First, compete with beauty. Beauty doesn't cost more than ugly. They were already spending a fortune, but those resources were being squandered, because the efforts were not designed to make the place full of life and art and culture.

The second strategy was to make the design of the city a collective project. The imagining of a city is one of the most powerful and important opportunities for citizens to experience and express their collective potential. And yet the design is often generated away from any public engagement. When I have been involved in change projects, I have seen what can happen when people discover the power they have to make their world. They move from consumer to producer, shifting from passive to active. They become designers, capable of engaging the world, questioning and challenging, and imagining possibilities.

The third strategy was perhaps the most radical. Design the nature of the city. Start with the natural systems and ecology of the place, and design our presence and long-term viability within that sustainable system. Sudbury has an amazing three hundred and thirty lakes inside the city limits. This offered a possibility to invert the conventional diagram of city and park: Usually we think of the park as a rectangle of green in an otherwise gray world, but in Sudbury we could imagine a park with a city in it—a park that happens to have a hundred thousand people living in it.

The fourth strategy was simply a statement: *Everything communicates.* It takes us back to the message that we are sending to young people. Everything we do, and the way that we do it, is part of that message. We can tell young, talented people that we want them to stay in Sudbury, but if our actions show that we don't care, nothing we say can matter.

There is enormous possibility in this situation. Redesigning Sudbury as a park with a city in it—and doing this as a collective effort could serve as a prototype of a sustainable northern city. And if we could accomplish this in a northern mining town, imagine the wider implications.

Some time after my presentation, a group of citizens contacted the studio to inquire about whether I would work with them to realize this vision. They called themselves Imagine Sudbury. How could I resist?

and said, 'I want to go to art school.' " The counselor told Mau that because he hadn't taken any art classes, it was too late. "I said, 'What do you mean it's too late? I'm fifteen years old—surely my fate can't be sealed!' So they let me enroll in a special art program downtown."

Mau stayed in high school an extra year, just taking art classes, and found a mentor in a sixty-five-year-old teacher, Jack Smith, who was in his last year before retirement. "I didn't know it at the time," Mau says, "but he was what Buckminster Fuller would call a 'comprehensivist'—he loved all the arts and sciences." Mau was soon doing his own color photography and printing, drawing, ceramics, filming, and typography. What hooked Mau on graphic design was the school's old one-color Heidelberg offset press; Mau refurbished it and used it to create his own four-color prints. "It was the most incredible year," he recalls. "I had complete freedom to work on whatever I wanted, always trying something new. Years later I was thinking about that experience and I realized that the whole idea of my work comes from that. It's like I've been re-creating that last year of high school in my studio ever since."

On the strength of those four-color prints made on the one-color press, Mau managed to get into Toronto's Ontario College of Art. He had never set foot in a city before going to the college in Toronto, but he wasted no time getting out of Sudbury. (Years later, after he'd become a well-known Canadian designer, his hometown would invite him back—they wanted to know if he could help redesign a dreary old mining town that couldn't keep its young people from moving away. Mau accepted the invitation and is now trying to transform the place of his youth.)

AFTER EAGERLY fleeing to Toronto, Mau didn't last long at the art college. He got into trouble for skipping his classes so that he could sit in on whatever other class happened to interest him more, including advanced courses. Then he dropped out altogether and drifted overseas, to London, where, owing to a connection and his impressive portfolio, he landed a job at Pentagram. One of the most prestigious international design firms, Pentagram represents a kind of creative nirvana in the design world. Formed as a collective by a group of the world's top designers, it has expanded greatly but continues to be run by a core group of star designers. People have spent the better part of their entire careers at Pentagram, never choosing to look elsewhere. As for Mau, he lasted a year and a half.

Mau says that while Pentagram did great work, the place felt too corporate to him. There was one other thing that bothered him. The firm's managing partners sat at desks that were at a higher level than the desks of the other designers. Describing this many years later in the kitchen of his home, Mau drew a picture to show a high desk hovering over a low desk. "It was like something out of Dickens," he said as he sketched.

Mau may have been extra sensitive to this because he'd become politicized in reaction to the Thatcher government, during the time of the Falklands War. He felt designers ought to be challenging government policies and practices that were, to Mau's view, unfair and even oppressive.

In this early stage of his career, Mau was already beginning to question whether he even wanted to be a graphic designer. When he was among designers who'd devoted their lives to the craft, he says, he couldn't help feeling that "they'd done a Faustian deal." He also comments that when he was putting in long hours in those days, "doing corporate stuff that had no meaning, I kind of felt as if I was designing my own prison."

By this time, there was a pattern emerging in Mau's life, and, having a designer's eye, he spotted it. He had a strong, maybe even irrational, resistance to being fenced in—whether in the town of Sudbury or in science or in a structured academic program, or in a corporate environment or even in a single discipline of design.

He returned to Toronto full of "stupid questions." *How can a designer make a difference in people's lives? Can design do more than just sell things on behalf of a client?* And above all, *How does one keep from getting bored?* Mau concluded, fairly quickly, that if he was going to find a design studio that was right for someone like him, he'd have to design it himself.

2. JUMP FENCES

How do designers connect, reinvent, and recombine? And what makes them think they can do all these things?

2.1 RECKLESS LEAPS AND "SMART RECOMBINATIONS"

The promise and the risk inherent in asking a stupid question—*How might something be done differently, or made better?*—is that the questioner may, in turn, be asked to provide an answer.

When that happens, a designer's search for new possibilities begins and sets in motion a process that is fairly standard. Design methodology generally involves conducting research (including, in some cases, deep digs into human needs and wants), creating sketches and models so that ideas can be shown to others and analyzed, and continually reworking and refining an idea or possibility, in response to feedback or new ideas.

The most creative, innovative designers follow these steps—but so, too, does just about everyone else in the field. So how do the more groundbreaking designers manage to separate themselves from the rest of the herd? They do it by jumping the fence.

Actually, they may jump a number of fences to get all the way to innovation, depending on the difficulty of what they're trying to achieve. When Yves Behar was asked to help design a laptop computer that could outperform anything on the market and do so at a quarter of the price, when Dean Kamen was enlisted to design a prosthetic arm with a hand that could pluck a dime off a tabletop or pinch a grape without crushing it, when Gordon Murray embarked upon the challenge of making the world's first wholly recyclable car . . . each of these designers was attempting to make the leap from the realm of known achievability (think of a field with a sign on the gate labeled WHAT WE KNOW IS POSSIBLE) to the much larger surrounding space (marked WHAT WE DON'T KNOW HOW TO DO YET).

Before making that leap, each of these designers started by envisioning something that did not exist. Then, each had to proceed in the face of evidence suggesting the thing *could not* exist, that it simply wasn't feasible. The breakthrough designs that resulted may be thought of as technological advancements, but the real leap was more of an imaginative one—because it took place, at least initially, as a glimmer moment in the mind of the designer. By relying on "abductive reasoning," or the ability to think about and picture what *might* be, designers can glimpse possibilities that lie on the other side of the fence.

That doesn't make the jump any less scary. Often, Dean Kamen has no idea, on his most challenging projects, whether he can clear that fence at all, or what kind of landing he's likely to have if he does. Kamen says he grapples with this uncertainty constantly: "You roll around in bed at night, thinking 'Am I trying to do something that's impossible?'"

All that tossing and turning notwithstanding, Kamen, like a lot of innovative designers, has an unshakable belief that he can alter reality, even in areas where he has no particular expertise. He didn't know anything much about prosthetics until quite recently, when he looked up from his desk one day to see several high-ranking military officers marching directly toward him. At the time, in early 2007, the U.S. military was attempting to deal with some of the fallout from a poorly designed project known as the Iraq War, which was sending home scores of young people with missing arms. The military men wanted Kamen to create a replacement arm with unprecedented mind-hand coordination.

Kamen knew enough about engineering to know that what they were asking couldn't be done. "I said to them, 'You guys are crazy.'"

But here is the interesting thing: While Kamen was saying no, the whole time he was thinking, "'Well, why not?' I figured I could get partway there."

About a year after Kamen's meeting with the generals, there was the arm in Kamen's studio, attached to a life-size replica mannequin of the Terminator. The arm had already been tested by an amputee at a live demonstration before top government and military officials in Washington—and, yes, it could indeed pick up a grape, or an egg, or a coin off a tabletop. It could also give you one hell of handshake.

THE REASON why Kamen and Behar believe they can do the impossible is partly because they've already done it. At least that's one of the theories of Roger Martin, dean of the Rotman School of Management. Martin, whose prestigious school is located in the vicinity of Mau's studio in Toronto, has analyzed the way designers think and create. Mau has been one of his study subjects, as have Milton Glaser and also the esteemed designer Massimo Vignelli, creator of the innovative New York subway map introduced in the 1970s. According to Martin, top designers have certain common characteristics, one being their rock-solid belief that reality is subject to change. When designers are confronted with a challenge that has no real-world answer, Martin notes, "instead of saying, 'Well, that's life,' they are inclined to say, 'No, there has to be a better answer out there if I think a little bit harder.'"

And once they've come up with a breakthrough solution on one problem, they're better prepared to do so the next time, and better still the time after that. Martin refers to this as the "upward spiral" of solving problems, wherein the more you do it, the more you can do it. An experienced designer will look at an "impossible" problem and his/her reaction, Martin says, "will be, 'Ah, I know this game. I've been in this situation before. It's fine if there are no existing good answers out there, because my job is to design a better answer.' And they proceed with confidence from there."

According to Martin, designers "live in an expansive world where they believe the only thing limiting us is the stuff we haven't figured out yet. And they're excited about it. You'll hear them say things like, 'I'm working on this really cool problem that has no answer!' That's what they live for."

To be comfortable taking on tough design challenges, you must embrace ambiguity and complexity. Most of us, Martin notes, try to simplify problems and make clear-cut choices; we strive to construct a single, clear "mental model" when we're thinking about a challenge and trying to envision changes and solutions. But a designer (and, Martin points out, this can apply to other creative minds, too, including some top business executives he has studied) is comfortable holding conflicting ideas in his/her head at the same time. "The designer lets a lot of different models float around in the mind at the same time. And they select parts and pieces from those existing models to create new and better models."

This is another way that designers "jump fences"—by mentally hopping from one realm of thought to an entirely separate one, connecting ideas from that first world to ideas from the other. These connections are actually problem-solving insights—*What if I put this with that? What if we were to take the sensor technology from the Segway and use it in an arm socket?*—that the designer and author John Thackara calls "smart recombinations." They often connect ideas and possibilities that would seem to be unrelated.

For example, when the designer Van Phillips was working on the challenge of creating a more flexible prosthetic leg some twenty years ago, he relied on a number of connections and recombinations. Foremost in his mind at the time was the letter "C"—his father owned a C-shaped Chinese sword and Phillips had always been impressed by the flexibility of the blade. At the same time, Phillips was studying the mechanics of diving boards; he wondered if the same kind of spring force could be applied to a prosthetic. Somewhere in the midst of all this, he happened to learn that the hind leg of a cheetah functions in a very distinctive way, with the tendons compressing and releasing in a manner that yields great elasticity.

Phillips began to connect all of this—along with his knowledge of the unique properties of carbon graphite—to begin to try to jump the fence to a better artificial leg. Phillips says his college professors at the time, upon learning what he was attempting to do, advised him not to waste his time. But Phillips pushed on, perhaps because he had a real stake in the outcome of his work: A few years earlier, when he was twenty-one, his left leg had been severed below the knee in a water-skiing accident. Phillips told the *New York Times* that he became "obsessed" with designing a replacement foot— creating hundreds of prototypes, baking them in his kitchen oven, then test-

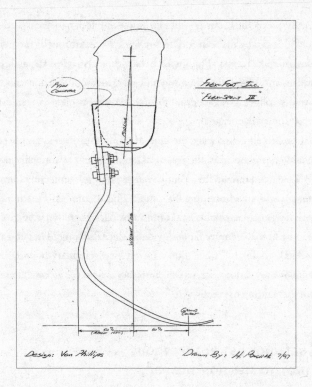

ing each foot himself. Often, the foot he'd made would quickly break under his body weight, sending Phillips tumbling to the ground; he'd pick himself up and begin work on the next iteration. The lower-leg and foot prosthesis he eventually created—appropriately named the Cheetah, with no heel and a distinctive C-shape—changed the prosthetics industry, as well as Phillips's own life, enabling him to run and engage in other sports. The Cheetah drew worldwide attention in 2008 when the track runner Oscar Pistorius competed for an Olympic berth, running on two Cheetahs and looking like a man from the future. But as Phillips knows, it could all be traced back to the time he connected bits and pieces from the animal kingdom, the local swimming pool, and the battlefields of ancient China.

All of us have the ability to make these surprising new connections—researchers say the brain's right hemisphere is fertile ground for such far-ranging, hopscotching activity—though innovative designers seem particularly adept at it. It can be the source of technological breakthroughs for an industrial designer like Kamen, Phillips, or Behar, but it also helps visual designers to connect and blend ideas in a way not seen before. Indeed, it cuts across creative disciplines as it is relevant to anyone trying to create something that is at least somewhat original.

Is it possible to encourage the kind of thought processes and behaviors that enable designers (and the rest of us) to jump fences and make connections? That's a question that Mau began exploring twenty years ago as he first set up shop, and one that he's still trying to answer. This much he's learned: To get to originality and innovation, designers must be given free range to be able to venture far and wide in their thinking. They mustn't be penned in by habit, by convention, or even by disciplinary boundaries—all of the many and various barriers that, in Mau's view, represent "attempts to control the wilding of creative life."

2.2 GETTING LOST IN THE WOODS (NOT TO MENTION THE FOG)

When Mau opened his Toronto design studio in the mid-1980s, it was a kind of experiment in which he was the main subject. He'd returned from London full of fire, politically engaged, and looking to change the world—which, for a graphic designer, usually means making posters for movements and causes. Mau started a small design firm called Public Good, which promoted union initiatives, educational programs, and arts performances. After a couple of years—he hadn't changed the world much in that time—Mau decided to accept an offer from a New York publisher to design a new book series.

Zone books would be a defining part of his work for the next two decades. It was an ideal job for Mau because the publisher wanted every new book in the series to be a fresh reinvention. Each Zone book delved into a single, complex topic (from an exploration of cities to an examination of cyberkinetics), and Mau was encouraged to change the format each time, to somehow match the subject. It was at Zone that Mau became a designer who cared as deeply about content as style. He worked closely with the writers and editors in an intensely collaborative free-for-all. "The idea with each book was to take a single important issue and have smart people come at it from all angles," Mau says.

Mau's design work on Zone was wildly experimental. In some ways, he was doing what he'd done back in his high school arts program—using whatever tools were at hand to try to create wholly original effects. On the first Zone book, about "the contemporary city," Mau sought to turn the book itself into "an urban object," with the qualities of a city—the friction, the complexity—expressed through the montages and the multilayering of Mau's design. The result looked high-tech but was all done by hand, in the late days of predigital design. Mau photocopied hundreds of images and then superimposed layer upon layer, "just by passing it through the copier machine over and over."

Zone put Mau's new studio, Bruce Mau Design, on the cultural radar. When architect Frank Gehry was looking for a book designer, he was introduced to Mau and took a shine to him. Then Gehry decided to offer Mau more than just a book assignment. At the time, in the early nineties, Gehry was building the Walt Disney Concert Hall in Los Angeles and thought it might be interesting to have Mau design the interior graphics.

"I'd never designed anything like that in my life and had no idea how to do it," Mau says. He later found himself in a similar situation when he was asked to help design a library in Seattle, working with architect Rem Koolhaas; and it happened again when he designed his first city park (Toronto's Downsview Park), and then his first biodiversity museum (actually, the world's first biodiversity museum, done with Gehry in Panama), and then his first bookstore redesign, done at the request of Indigo, the Canadian chain. In the case of the latter, the fact that Mau didn't know how one is supposed to design a bookstore only gave him more freedom to try to do something his own way. What resulted was a bookstore designed and arranged like a book: A visitor could effectively "flip" through its table of contents by scanning the in-store signage.

On these various projects, as Mau and his small staff of designers found themselves in unfamiliar territory, Mau learned a couple of lessons. "We figured out that, no matter what the project is, we know how to design—and that is a universal methodology." The other lesson: Feeling lost on a project can be the first step toward finding an original solution.

IN FACT, Mau began to think of that period of "not knowing" at the outset of an unfamiliar task or challenge as a window of opportunity. "When you don't know what should be done, or how something is supposed to work, it's a brief pocket of possibility," he says. "You're free to speculate on something unencumbered by the conventional structures."

Mau developed a working model that encouraged the designers in his studio to speculate first, and research later. The point was to come up with wild ideas, scenarios, possible solutions; then to sketch them, film them, express them in the form of collage pictures cut out of magazines, tape them to the wall. But the idea was *not* to research them to a premature death. Mau came to believe that in those earliest stages of thinking about a problem, when people were unencumbered by data and expert opinion and conventional wisdom, they were most likely to happen upon and be open to fresh, unusual, and possibly game-changing ideas. "If you start out speculating," he says, "you'll find yourself saying, 'It'd be exciting to try something like this—I wonder if it's possible?' Then later you do the research to find out."

During his periods of feeling "lost" on projects, Mau also observed a phenomenon that reinforced his beliefs. There was a unique kind of creative energy and resourcefulness that seemed to come to the fore when people were on unfamiliar turf. This led Mau to his "lost in the woods" theory, which he articulated in one of the journals he produced for internal use in the studio. "If you're lost in the woods," Mau wrote, "everything about your surroundings takes on added significance. Suddenly you have to navigate and negotiate every detail of the environment, processing all of it while trying to regain your bearings." When people are in this "hyperattuned" state, Mau reasoned, it's an ideal time to experiment and speculate, because the mind is wide open and the senses are alive.

Mau's "woods" theory is closely connected to another analogy he uses, one that he borrowed from the military. In "the fog of war" soldiers often must make quick, instinctive decisions based on limited and degraded information. Because of the difficult conditions and uncertainties of the battlefield, they may only have 20 percent of the information they'd ideally want in order to make a fully informed decision. But the trouble is, "if you're on the battlefield and you wait until you have 100 percent information," Mau says, "by that time, you're dead."

The commanders who do well in the fog of war are those who learn to see through the haze, recognize the important pieces of information that are available, connect that with experiences from other battles, and, ultimately, trust their instincts. They do all of this under intense life-and-death pressure, because they have no choice. But it can result in optimal performance and enhanced creativity.

Mau felt the design process could benefit from that, as well. Of course, there were no life-and-death circumstances for his designers. But in terms of the fate of ideas, Mau found that certain of them—usually the more tangential, unexpected ones—might only come to life during the civilian equivalent of the fog of war, in that brief window wherein people were lacking information and therefore were less likely to prejudge. In this scenario, knowledge becomes the enemy.

MAU'S "FOG OF WAR" approach runs counter to the way most research-driven companies behave, as well as to our culture of experts. The addiction to preliminary research has tended to yield analytical data so dense and difficult to navigate that it forms its own kind of fog. Many design firms, too, are grounded in heavy up-front research. In fact, there is something of a schism in the design world now. The "research first" crowd points to how creative and innovative design research has become, and indeed it's true that the best design research is itself designed to produce original and sometimes startling insights.

That said, there is a growing faction, both inside and outside the design world, that argues in favor of approaching a problem or challenge with less expert knowledge, not more. Innovation expert and author Cynthia Barton Rabe has helped popularize the notion that breakthrough ideas are more likely to come from "Zero Gravity Thinkers," meaning those who aren't weighed down by expertise and conventional wisdom. Some innovation pundits now refer to the "curse of knowledge," which holds that as expertise increases, creativity tapers off.

To maximize creative opportunities during the temporary state of not knowing, Mau developed a number of studio guidelines focused on ways to encourage experimentation and free association. Mau felt it wasn't enough just to preach the value of experimentation. There was a need to provide people the time and security to experiment, to connect ideas and explore adjacencies—in a word, to *drift*. And giving designers an hour or two to drift is not enough, Mau believes. That limited amount of time will bring forth the surface ideas, but not the "deep woods" ones. In the studio, Mau tried to set project schedules so that people might have days or weeks to drift. And during this period, criticism of ideas was to be tempered if not withheld. "Most people are too quick to criticize and cut off ideas," Mau says. That criticism should be postponed until the later stages of creative development, when all ideas get subjected to rigorous critical analysis and testing to separate out the best. One of the laws of Mau's studio is: *Harvest ideas. Edit applications.*

Mau also has tried to foster experimentation by emphasizing its value. What often keeps people from experimenting is the notion that if the effort doesn't yield something immediately usable—and the reality is, most experiments don't—then the experiment itself has been a waste of time. But in design, as in science, an experiment can have great importance regardless of outcome. It can better guide subsequent efforts—so that the failed effort becomes an important step on the path to producing a successful design down the road. With this in mind, IDEO and other leading design firms have embraced the "fail early, fail often" model, while Mau, for his part, has instilled in his studio the practice of "capturing accidents," wherein failed experiments are documented, preserved, and practically worshiped. They are viewed as successes that simply have not happened yet—or, as Mau puts it, each one is "the right answer in search of a different question."

Mau thinks that, as a culture, we judge experiments much too harshly and, as a result, it limits our progress. Ambitious efforts to create change— such as Dean Kamen's Segway or Yves Behar's XO laptop—are sometimes held to an immediate pass/fail standard: If they don't change the world overnight, they're marked F. But in design, and particularly in complex design that seeks to alter the way we live, success is likely to arrive in stages, via a series of experiments or "iterations." The first version of a radical design to make its appearance in the world may be imperfect, but it still can play an important role. "It shows that a different and maybe better way of doing things is possible," Mau says, "and after that, nobody can claim it's impossible anymore."

2.3 IT'S AMAZING, IT'S CUTE, IT'S GREEN. COULD IT BE A BABY DINOSAUR?

When Mau's Massive Change show was in Chicago in the fall of 2006, of all the many intriguing creations on display, perhaps none drew more admiring looks than the prototype version of the small green laptop computer designed by a team that included Yves Behar. At the time, this "hundred-dollar laptop," as it was first known, had not yet been unveiled to the wider public. It looked like a toy (which was the designer's intention) with a rubberized lime green skin. The thing just made you want to handle it and possibly even taste it.

Mau says it achieved exactly what design, at its best, can provide: a delightful and simple solution to a profound and far-reaching challenge (the challenge here being, How do we begin to bridge the world's great digital and education divide?). Of course, in the two years that followed, the XO laptop would garner much attention as it was initially hailed as a design wonder and then, gradually, dismissed by some as a major disappointment. Today, depending upon whom you talk to, it represents a small miracle or a promise unfulfilled.

But there is no disputing that Behar and the One Laptop Per Child (OLPC) team, headed by Nicholas Negroponte of the Massachusetts Institute of Technology, managed to jump the fence in terms of going where no computer maker had gone before. Which raises the question: Why is it that a small ad hoc team, headed by an academic in Negroponte and a tech-industry outsider in Behar, was able to design something that was beyond the capabilities of the major computer companies and extensive R & D labs?

The simplest answer is that Negroponte's team believed in the goal and actually wanted to do it; others did not. Dean Kamen has an interesting take on this. "Right now, the capabilities are out there to do just about anything we want to do," he says. "We used to ask ourselves, 'What can we do?' But now the real question is, 'What should we do?'" As Kamen sees it, some people believe we should use all available capabilities to tackle major problems and try to achieve progress. Others are content to pursue more modest and conservative goals. "Like a cell phone that comes in a new color—that's some people's idea of innovation," Kamen says.

For his part, Negroponte was thinking big—maybe too big. He wanted to make laptop computers widely available to poor children around the world and felt that in order to do so he'd have to hit the magical hundred-dollar price point, so that countries could afford to buy the units in bulk and then distribute them to schools. But he did not intend to make a "cheap" computer. Negroponte saw the available options as two equations:

1) cheap components + cheap labor + cheap design = cheap laptop
2) extreme integration + advanced manufacturing + large number + cool design = inexpensive laptop

In pursuing the second equation, he assembled a few top-notch engineers and technologists from Silicon Valley and also enlisted a leading design firm, Continuum, to work up some early plans and prototypes. One of Negroponte's boldest moves was bringing in Behar.

A Swiss-born man-child with an angelic face and a mop of curly dark blond hair, Behar works out of a small industrial design shop named fuseproject, housed in a narrow two-story building that looks like a corrugated metal box, located on a gritty block under the highway on San Francisco's east side. A couple dozen designers work inside the modest building. Toward the back is a snug tool room and a workshop that looks to be big enough for a few elves.

Before Negroponte and OLPC came calling, Behar had dabbled in computer design early in his career, though he later became known for his work on more stylish products—a lamp that resembled a blade of grass for Herman Miller, eco-shoes for Birkenstock, sleek Jawbone headphones—all designed in innovative forms that earned kudos and made Behar much in demand in corporate America. He says he almost didn't take on the laptop project because he felt that computer makers had little appreciation for design. During his brief early time in Silicon Valley, he was turned off by tech nerds who used *Star Wars* lingo to code-name their supersecret projects. "Meanwhile, I'd be the one raising a hand and asking stupid questions like, 'So, what's the Num Lock key for?'" Behar eventually concluded that when it came to computer design, "all anyone cared about was how many new features they could cram into the computer."

Despite his misgivings about Silicon Valley, Behar was impressed with Negroponte's OLPC group and its bold change-the-world mission. "Nick was an educator and a tech guy, but he understood that design was going to be the way that kids would connect with this device," he says. Plus, Behar adds, "Nick said that he was going to get rid of the Num Lock key."

The hundred-dollar laptop project involved a series of interrelated challenges, all of them formidable, some seemingly insurmountable. For one thing, the device had to be simpler, from a functional standpoint, than any computer out there, because it was going to be aimed at children who'd probably never used nor even seen a computer and who might have no one to teach them how to use it. This meant that the design had to pay extra attention to intuitive design techniques such as "mapping," using universal icons to make it clear how to navigate the computer's functions, and building in lots of forgiveness, so that if you made mistakes while playing with the computer, it was no big deal. The laptop needed to be built like a miniature tank, able to withstand harsh outdoor conditions including sun and sand, as well as rough handling. It would have to be powered and recharged without using electricity, which is unavailable in places where the computer might typically be used. And above all, it had to be cheap—about one-fifth of what the lowest-priced laptops cost at the time.

WHEN WORD circulated through Silicon Valley that this miracle machine was in the works, the effort was publicly dismissed and even ridiculed by executives at Intel and Microsoft: There was no way to make a laptop for a fraction of the cost of the cheapest one made (never mind powering it without electricity!).

Negroponte had an ambitious goal to get the computers to market quickly, so there wasn't much time for up-front research. "We did speak with some educators in developing nations and talked about some of the challenges in that environment, but otherwise this was about trusting the designers' intuition," says Kevin Young of Continuum, the firm that did some of the initial design work on the laptop.

At those early stages, the designers clearly followed the Mau approach of "harvesting ideas." A wide variety of radical formats were proposed, including the notions that the computer might take the form of an integrated backpack that got its energy from the sun or, better yet, that it might derive its energy from having a power source strapped to the leg of a cow.

As it came time to edit ideas, some of the wilder ones were deemed impractical or too fanciful. Designers sometimes must part with interesting ideas early on, lest they become too attached to something that is, for real-world purposes, a "dinosaur baby" (a term coined by IDEO designer Paul

Sketch by Yves Behar

Bennett to describe a quirky and idiosyncratic design creation that is destined to be loved only by its creator).

By the time Behar took over from the Continuum designers, it had been decided that the computer would take the more conventional form of a notebook, but that was about the only conventional aspect. For instance, this would be the first computer with ears. Negroponte and the early designers had figured that if they put antennae on the laptop, they could create a "mesh" network that would connect the kids' computers with each other and with the Internet, even in remote areas with no satellites or towers. Behar took what could have been a design negative—floppy, ugly, obtrusive antennae—and turned it into a distinctive feature. He made the ears look like part of a cute face and designed them so they'd flip out of the way when not needed.

Behar also took one of Continuum's early blue-sky ideas—the crazy notion that an old-fashioned hand crank might be used to recharge the laptop computer battery—and made it a reality, along with an alternative yo-yo-style pull cord; either of these would allow the computer to be powered manually by a child when there was no other power source available. Behar also put in a solar panel as yet another power source. Not that the device would end up requiring much power—it ran on about one-tenth the juice of other laptops, which made it green in more ways than one.

The OLPC team zeroed in on the most costly part of a laptop computer, the illuminated screen, by experimenting with prototypes using various screen materials that might take advantage of natural sunlight. A key breakthrough came from one of Negroponte's crack Silicon Valley technologists (the collaboration gave Behar access to tech expertise he didn't have), who managed to reduce screen costs by leveraging backlighting. The final screen had five times the resolution of other laptops, and could be read in direct sunlight.

Perhaps the biggest barrier—producing the laptop at a fraction of the cost of competitors—wasn't quite as daunting as it first seemed. While the conventional wisdom held that laptops cost several hundred dollars to produce, the truth was that typically half of that amount went into marketing and other external costs. When you got down to the nitty-gritty of making the computer, the cost was much lower and could be trimmed even more by streamlining and eliminating all dispensable parts. Behar ditched the internal fan because he figured out how to rearrange the computer's structure so that it would disperse its own heat. And he made every little part do double or triple duty: The ear antennae, for example, also doubled as bumpers to cushion a fall. And when you closed the computer, the ears slid in place to form a protective cover for the ports.

With all of these efficiencies, the cost came down to $188, which was as low as Behar and the design team felt they could go without too much "satisficing" (which refers to making design compromises to keep the cost down). Behar and Negroponte did not want to make something that would be perceived as cheap, though in fact their laptop, even at the higher price level, was about one-third the cost of a bargain-basement-cheap laptop.

For greater durability, Behar encased the computer in rubber with a goose bump texture (Behar is practically obsessed with tactility, believing that if you can get people to want to touch something, the battle is half won). And while he wanted all the computers to be that cute lime green color, he also wanted each kid to feel that his/her computer was unique, so he used twenty different colors for the X and the O in the big XO label on the computer, resulting in four hundred color combinations—which made the odds pretty decent that you'd have a different one than any of your closest friends.

When Behar's prototype was presented to the MIT panel heading up the project, the room broke out in applause. And just a few months later, his prototype was the talk of the 2007 Consumer Electronics Show in Las Vegas. Executives at Intel and Microsoft continued to dismiss Behar's creation. But then, suddenly, they announced that they, too, were going to begin work on a low-cost laptop. Because now, of course, such a thing was entirely doable. The realities of the laptop world had changed overnight. Yves Behar had jumped the fence. What Behar didn't know at the time was that there were other hurdles still ahead, and they would prove harder to clear.

2.4 THE FISH LIGHT, AND OTHER LATERAL THOUGHTS

Behar says he never worried for a moment that he would be unable to do all those things that had never been done on a computer before. Not that there weren't sticky points along the way—trying to figure out what to do with the battery almost drove him batty, he confesses—but he always knew he'd figure it out. The capabilities were all available, Behar points out; they just needed to be put together in the right way. Behar and his fellow designers did achieve some technological advances, such as with screen materials, but for the most part the product was designed by connecting existing pieces and parts in a smarter fashion, eschewing anything that was unnecessary, and combining fresh approaches with some very old ones (that hand crank looked like something from the Flintstones).

This brings us back to John Thackara's "smart recombinations" idea and the significance of it. Designers such as Kamen and Behar aren't really inventors in the sense of making things from scratch. As Kamen says, "I take what's already out there and figure out how to use it in a new way." If you think of innovative design this way—as a method of cobbling together ideas, influences, and resources that already exist—it makes the act of invention slightly less daunting. "Designers are needlessly constrained by the myth that everything they do must be a unique and creative act," according to Thackara. What they really need, he says, is "the capacity to think across boundaries," "to draw relevant analogies," and to "put old knowledge into a new context."

Not to suggest that that's easy to do. The best designers seem to have a natural eye for spotting patterns and discerning possible relationships between things that most of us view as being separate and unrelated. Once they see a possible relationship, they work to make the pieces fit. "Designers are trained to synthesize," according to Charles Cannon, a professor at the Rhode Island School of Design.

If the designer is T-shaped—with broadened interests that go beyond a single design specialty—it only helps, by providing more material and more points of reference to draw upon when seeking to create new connections. Sometimes that material actually is material—industrial designers like Behar hoard scraps in their workshops and file drawers, saving a little piece of everything they've ever used on projects in hopes they can use it again on something new.

But the bits and pieces often reside in the mind of the designer. Michael Bierut's brain "is a compendium," says his Pentagram partner Paula Scher. "He absorbs everything and then uses what he needs at the right moment." Bierut himself compares design to "doing a crossword puzzle"—you have to fill in the blanks with the right idea, which may be based on something you saw or learned years earlier.

(On the facing page is what the puzzle of designing the Cheetah artificial leg may have looked like to Van Phillips.)

George Lois speaks of the thunderbolt that claps when a designer makes a connection. But this bolt "does not come from heaven," Lois says. "It may seem like it's from out of the blue, but it's all from life experience, from your understanding of the world around you, of history, of art, of sports. You have to have a real sense of what the hell's going on in the world and in the culture."

		¹S	P	R	²I	N	³G			⁴G
		W			T		R			U
	⁵F	O			H		A		⁶F	I	T
⁷B	O	A	R	D			P		L	
	O	D			⁸N	O	H	E	E	L
	T	O			K		I		X	
		F			I		T		I	
⁹R	E	A	C	H		¹⁰C	H	E	¹¹E	T	A	¹²H
U		H			A			N		I	
¹³N	A	D	I	R		N		A		N	
		N						C		D	
		¹⁴E	L	A	S	T	I	C	I	T	Y

Design by way of filling in the blanks: After designer Van Phillips lost his left leg water-skiing, he became obsessed with creating an improved artificial leg and **<5-DOWN>**. He relied on a series of mental connections and "smart recombinations," beginning with his interest in a C-shaped **<1-DOWN>** whose blade had extraordinary **<14-ACROSS>**. He connected that with the mechanics of a diving **<7-ACROSS>**, which derives its power from **<1-ACROSS>** force. Turning to the animal kingdom, he was inspired by the amazing **<10-ACROSS>**, whose **<12-DOWN>** leg demonstrated unique contracting qualities each time the animal would **<6-DOWN>**. Phillips realized that the strong-yet-flexible material known as carbon **<3-DOWN>** might allow him to emulate these qualities. He hit a **<13-ACROSS>** when his professors looked at his plan and said, "No way you can do it." Phillips responded with the designer's mantra: **"<2-DOWN>."** He trusted his **<4-DOWN>** instinct, which told him it was possible to **<9-ACROSS>** a new level of prosthetic design and that it was up to him to find a way to **<11-DOWN>** his plan. The artificial leg that Phillips eventually designed, named the Cheetah, has a distinctive C-shape because the foot has **<8-ACROSS>**. The Cheetah drew acclaim during the 2008 Olympic trials, as double-amputee Oscar Pistorius showed he could **<9-DOWN>** faster than anyone imagined possible, using two Cheetahs. Phillips himself uses one as he runs each day to stay **<6-ACROSS>**.

EVEN AS they tap into a wide range of influences, designers trying to think of fresh ideas tend to have the same problem as the rest of us—their minds too often gravitate naturally toward the familiar. It's true even for someone like Stefan Sagmeister, considered one of the most wildly original and unconventional designers working today (he has designed everything from rock posters to three-dimensional displays on the meaning of life, one of which was made entirely from ripening bananas). Sagmeister, an Austrian native whose design studio is based in New York, says it's a constant struggle to avoid falling back on familiar ideas.

In discussing this, Sagmeister cited the philosopher Edward de Bono's theories about "lateral thinking." De Bono found that the brain has a natural propensity for repetition because that's when it functions most smoothly and efficiently (it's why you can multitask easily as long as the tasks you're doing are familiar). "But when it comes to ideation, or thinking of something completely new, this whole repetition thing is actually a drag," Sagmeister says. Wanting to do the same things over and over, the brain will tend to think of ideas it has thought of before, as if they were new. "Happens to me all the time," Sagmeister says. "I want to think about a new idea and the first thing that comes up is, 'Oh, can we do a die-cut like we did three weeks ago?' What de Bono says is that people build their whole careers on this kind of repetition."

To avoid that, Sagmeister and others go to great lengths to try to force themselves to "think laterally"—to break out of the familiar patterns and push their thoughts down unfamiliar pathways, sometimes by trying to make illogical connections. "It can be helpful to think about an idea from a point of view that makes no sense whatsoever," Sagmeister says.

"Let's try this out," he continues. "Say I have to design a new ceiling light. And so now I look around and the first thing I see is a fish." (Sagmeister has a fish mounted on the wall in his office.) "So now the idea is, think of that ceiling light from a fish point of view. "So, okay, can that light be something that you catch? Or maybe it has scales, this light. That could be nice. I recently swam through silverfish in Venezuela—maybe we could have a ceiling light that is a thousand silverfish that are so thin. Yes, this is an idea, actually. If we would have extremely thin silver scales, so thin that they move by heat, and if we have a hot light in the middle and a lot of super-thin scales around it, and we put that light on, all the scales would move toward the heat. Actually that is not a bad idea, and I'm going to write that down."

Sagmeister's trick of coming at a challenge sideways is slightly less extreme than an approach made famous a couple of decades ago by the designer Tibor Kalman, who was a proponent of doing things "the wrong way." Kalman was known for purposely creating images that were upside down or in some way jarring; it was Kalman's way of breaking the mold.

The approach of trying to view design challenges from unusual angles can be effective because "it gets you looking in the opposite direction from everyone else," says Tom Monahan, whose Providence-based creative coaching firm, Before & After, works with designers and marketing executives on how to think more creatively. Monahan uses an exercise known as "180-degree thinking" in which people start out trying to conceive of something that would have the opposite effect of what they're actually trying to create—such as a car that is unable to move or an oven that doesn't cook. "You start out making something badly, and then see if you can make that bad thing into something good," Monahan says. "Along the way, you may happen upon some unusual ideas and connections."

According to Monahan, it's important to "disengage"—to break the pattern of the way you might normally think about solving a problem. Several designers said they also think it's important to disengage in the physical sense when trying to come up with ideas. Sagmeister is a believer in taking creative sabbaticals or, at the very least, unplugging and disconnecting. "If things are coming at you—phone calls, e-mail, co-workers—you're constantly reacting, and in fact it's easier to react than to create. So all of us who constantly complain about e-mail—in that complaint is also an excuse, because it's inherently easier to return e-mails than to actually create something."

Recent studies on insights, conducted by Northwestern University professor Mark Jung-Beeman, have found that these mental glimmer moments occur most often when you temporarily disengage from a problem and ease up a bit. If you focus too intensely on the problem, you tend to get stuck in the more logical left hemisphere, but as you relax, the cortex is freed up to conduct a more far-reaching search through the right hemisphere of the brain. What it's looking for are remote associations that can help solve the problem in unusual and perhaps illogical ways. According to Jung-Beeman, when a "serendipitous connection" is made, the insight suddenly becomes clear.

Bruce Mau understood the value of disengagement before it was documented in these studies. Through the years he has urged designers in the studio to "slow down" and to try to "desynchronize from standard time frames" in order to allow the creative mind to work its magic. He believes creators must step away from the work frequently. But he also believes that they should "stay up late"—in fact, that is one of the laws in his manifesto.

On this point, Mau writes: "Strange things happen when you've been up too late, worked too hard, and you're separated from the rest of the world."

2.5 REENGAGING WITH REALITY

So: What happens after you disengage, stay up late, get lost in the woods (not to mention the fog), drift for a while, look at things the wrong way, come up with a fresh connection, and, ultimately, jump the fence? A designer with an original idea is just getting started. He/she must bring the idea to life in a preliminary form that can rally support behind it. And more often than not, the designer must continually refine and improve the idea, based on ongoing research and feedback. Even when the idea might seem to be in finished form, the work may be just getting started—particularly if a design represents some form of radical change, a true jump over the fence. In these instances, it may be necessary to design, as much as possible, how the outside world will receive and engage with this startling new creation. It was at this stage in the design process that Behar's little green offspring began to take on the look of a baby dinosaur.

As Mau has noted, sometimes when you're trying to solve one design problem, it becomes clear that you really must solve ten, or perhaps a hundred. Behar and Negroponte thought they needed to make a cheap computer that would be easy for kids to use, and they did that. But they didn't factor in how difficult it would be to get that computer out there in the world. As it turned out, they had to deal with competitors who tried to undermine their efforts, with educational bureaucracies that were resistant to change, and with cultural barriers that included strongly differing ideas on how children should be educated.

The laptop was designed by OLPC so that kids could basically teach themselves, using an open-source platform and a network that would allow them to connect easily to other kids' computers. But in a number of countries, particularly in Asia, educators weren't keen on the idea of ceding authority and educational freedom to students; this was seen as more of a Western approach to schooling. There were also concerns in some countries about who would provide the training and technical support needed. Some were spooked by the fact that the laptop program was using the Linux operating system, instead of the more familiar (and therefore presumably safer and more reliable) Microsoft system. It didn't help that Intel—which at one point signed on as a supporter and partner of the OLPC effort—began to try to lure away customers who'd signed up for the laptop program by pitching its own, more expensive version of the product.

Clearly, Negroponte's group made its share of mistakes—starting with extremely rosy early projections. Negroponte had predicted the group would have 150 million laptops distributed to kids around the world by 2009, but by fall of 2008 OLPC hadn't yet cracked the one-million mark. There was group infighting over whether to yield to pressures to offer the Microsoft operating system on the computer (Negroponte eventually agreed to do so), and there were also delays in production and shipping.

ALL OF these problems notwithstanding, it was still surprising when the media knives began to come out in mid-2008. It was no longer just cutthroat competitors "dissing" OLPC. Now, suddenly, pundits who'd initially lauded the laptop effort were beginning to call the whole thing a flop. Bruce Nussbaum, the highly influential innovation and design editor for *Business Week*, wrote what amounted to an obit for OLPC on his blog, headlined: "The End of the One Laptop Per Child Experiment: When Innovation Fails." A big *Business Week* cover story followed, containing a critic's charges that the laptop group was "arrogant" and that they even might be (*gasp!*) "cultural imperialists."

Immediately after these articles broke, the normally cool, laid-back Yves Behar looked like he might blow a gasket at a design fair in New York. "What do they expect? Of course there's going to be resistance," he snarled. "We're out to change the world!" Never mind that that was the sort of grandiose utterance that might cause people to want to pinprick OLPC in the first place; the point is, Behar had a point. Why did there seem to be a rush to write off and dismiss an experiment that, really, was in its early stages, perhaps just its first iteration?

Mau shared Behar's irritation. "What OLPC did was the farthest thing from a failure," he says. "They changed the game. Never again will computer companies be able to claim you can't do a laptop for under $200." And as Negroponte notes, the small netbooks that have surged into the market in the past year feature streamlined design similar to the XO's. Still the problem for the XO, as Mau sees it, is that when it comes to ambitious experiments and efforts to bring about change, "people seem to focus much more on what doesn't work than what does work."

Dean Kamen concurs. Furthermore, he believes it's a symptom of a society that has become averse to and afraid of real change, the kind that is often messy. "Think about the kind of big public efforts we used to do—putting in subway systems and building airports, practically overnight—and now look at us today," Kamen says. "We're not willing to do the hard work and take the risks that are part of making progress." Kamen says that to bring about change, "you have to kiss a lot of frogs." You must try things that are hard and that may not work right away (or ever). He thinks that in today's world—or in the Western world, at least—we're losing our taste for frogs.

Mau doesn't quite go along with that. He believes the will to create change is actually quite strong and growing these days among individuals like Behar, Negroponte, Kamen, and countless other innovators and would-be innovators out there. But it's mostly coming from the ground up, he says. Large organizations and governments are resistant and must be brought along, sometimes kicking and screaming, by entrepreneurs, activists, and dreamers.

"It's not easy to do," Mau says. "When you're looking to do something truly different, you're out there on a limb. You can feel the branch moving under you in the wind. Sometimes you hear the sound of a saw. But you have to try to bring others out there on that limb with you. You have to say, 'Come, look what I see—look at the view from here!'"

————

BACK AT the design fair in New York, Behar initially seemed dejected because of the negative press (a dejected designer: that certainly did not sound promising). But his mood changed quickly when the talk turned to a still-secret upcoming project. He was so excited about it that he had to spill the beans. He found a quiet little corner on the show floor and pulled out a sketch pad and began to draw an image—the next generation of the XO laptop. "It's going to be amazing," Behar said as he drew.

It sure looked that way in his sketches: The little green notebook was going to morph from a book to a tablet. Kids could gather around and interface with it from all sides. It would be smaller, lighter, yet infinitely more capable. And the goal in terms of the price: seventy-five dollars.

So there it was: a new fence, even taller than the last. And Behar was already poised for the jump.

3. MAKE HOPE VISIBLE

The importance of picturing possibilities and drawing conclusions

3.1 THE VIEW FROM OUT ON THE BRANCH

A few years ago, Mau was asked by Tulane University in New Orleans to help with the design of a river museum. Almost immediately at the time, Mau began to think of the project in more expansive terms, as is his wont.

"We had this idea," Mau recalls, as he grabbed a pen and reached for a sheet of blank paper. "It had to do with using the museum as a way to open up a part of the university to people who are otherwise cut off from the college experience."

On the paper he started to sketch something.

"Do you know what's the single biggest indicator of a person's academic success?" he asked. "Your zip code."

He stopped sketching. In the drawing, a stick figure man was looking up at a high cliff, from whence another figure looked down on him.

"This is the typical young guy in New Orleans," Mau said, pointing with his pen to the figure at the bottom of the cliff. "The university system has left him behind. He's way down here—and college is way up there. How's he going to get up there? What's he going to do, jump? But if this guy is just left down there, we're looking at a nasty version of the future. Because if he's stuck at the bottom forever, he is not going to be happy. You don't want to find yourself on the street with that guy."

Mau resumed drawing. Now he sketched a ramp that zigzagged up at several levels, providing a gradual ascent to the top of the cliff. This was what he had in mind with his experiment at Tulane: He wanted to design the museum with various levels of participation so that someone coming in off the street could gradually get more and more involved. At the first level, you visited the museum just like anyone else; no commitment necessary. But you'd be offered a glimpse of the next level of participation, which, if you pursued it, would feature more interactivity and would begin to teach river science. At the next level above that, you would begin to participate in longer-term projects (river cleanups, counting turtles, etc.), sponsored by Tulane.

"So now you're accessing the university, you're part of the higher education system," Mau said. If you continued on to the very top level, you could attain the status of being one of the museum's "scientists in residence," while gaining accreditation with the university. "The key is, at every level you could see the opportunity at the next level," he said. "And you would see a clear path to get there."

As it turned out, Mau never had a chance to build his ramp. Hurricane Katrina happened—and washed away the river museum plan along with

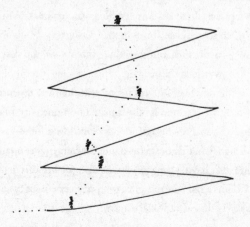

much else (though Mau is still working on ways to redesign the American university system, currently at Arizona State).

But after hearing this story and looking at Mau's sketches, what lingers in the mind is the possibility of that ramp: Somehow, it makes the idea of a more accessible higher education seem so feasible. And what's also hard to forget is that small, lonely sketched figure at the bottom of the cliff. Without the ramp, would he be left down there? What was to become of him?

IF THE design process starts with questioning what currently exists and then progresses to the next stage of seeking out fresh possibilities, at some point the designer must begin to communicate those new possibilities to others. As Mau previously noted, a designer seeking to innovate is in the position of being "far out on the branch"—and must convince others that they, too, should venture out there.

To do that, designers must show (or at least provide a glimpse of) what's visible from out there on the edge of that branch. "Everybody has ideas about how to fix things and change the world," according to the designer and former Apple creative director Clement Mok. "The difference with designers is that they have the ability to give form to those ideas."

That form need not be a thing of beauty. It can be as rudimentary as a freehand sketch on a napkin or as crude as a prototype made from foam rubber. Just the fact that there is a form at all is what matters. When a designer gives shape to an idea—even if the shape is rough and only temporary—it begins to transform the idea into something that is real, and that cannot be as easily ignored as words in the air.

It's one of the designer's chief talents: the ability to construct a dream, quickly. "A designer has one foot in imagination and one in craft," notes Brian Collins, who teaches design at New York's School of Visual Arts and also runs a design firm that bears his name. By being able to envision "what might be" and also having the capacity to draw or build representations of that vision, Collins says, "the designer can put a version of the future in your hands and ask, 'Is this what we all want?'"

Collins offers up an interesting definition of design: It is *hope made visible*, he says. That emphasis on the word visible doesn't mean he thinks design is all about appearances. On the contrary, Collins, like Mau, believes that meaningful design transcends aesthetics and is more about action—about reinventions and transformations. But those actions, Collins explains, are often rooted in some type of visual presentation of an idea that puts it in front of people and helps to rally their support. By making hope visible, Collins says, designers and innovators can generate momentum for all kinds of initiatives—from the development of new devices or services to the launch of social movements.

Mau has a slightly different take on the Collins credo about visibility. He thinks it's also important to emphasize clarity. This raises the ante a bit by challenging designers not just to bring future possibilities into view but to do so in a way that's compelling, accessible, and—above all—*clear*. "I have an obsessive killer's approach to clarity," Mau says, "in that I will find a way to get to it, no matter the obstacles." Mau believes that clarity is where the real transformational power of visual design lies. Aesthetics may get the glossy magazine covers, but when it comes to communicating change and progress, it's not so much about making things pretty as making things plain.

Achieving clarity is a considerable challenge, but it can help bring important change to business, to the way people live, or even, as we'll see in this chapter, to the medicine cabinet. It requires designers to find ingenious ways to simplify, organize, emphasize, and juxtapose. But often, it all starts with the basic act of drawing a picture.

SKETCHING IS "the archetypal activity of designers, and has been for centuries," according to Bill Buxton, a principal researcher at Microsoft Labs and a veteran designer himself. Buxton has studied and written on the significance of the humble sketch—not only as a tool for designers but as a critical element of innovation. To those who think of drawing as mere doodling, this may be surprising. But as Buxton points out, sketching can be central to innovation because it allows for fast and freeform exploration of multiple ideas. A sketch is easy to make (even if you're not a trained designer or artist; stick figures are generally sufficient), easy to understand, and easy to change. Thus the practice of sketching lends itself to collaboration and experimentation.

Interestingly, designers sometimes use the process of sketching not just to depict and share ideas, but to actually find them. To watch Milton Glaser sketch is to see a mind roam and search, via the point of a pencil. This is a phenomenon known as "exploratory sketching," and Glaser described it as follows: "There is a dialectic that exists between sketching and the way the brain functions, between the hand and the mind," he says. "When you're searching for an idea, very often what you'll do is create a kind of ambiguous sketch of it. And the brain looks at that and says, 'Ah, it could be this or that,' and then the hand transmits what the brain has observed and makes the sketch less fuzzy. Then the brain says, 'But maybe it should be like this,' and the hand accommodates again—and this conversation between the hand and the brain results in the development of an idea."

Glaser feels that he doesn't so much direct the process of sketching as go along for the ride. "You're sort of seeing where things go, as the brain responds to suggestions appearing on the page and transmits new information to the hand in a kind of feedback loop. You often don't know what you're doing until after you've done it."

Some designers find that building very simple three-dimensional models—often put together like makeshift toys, from whatever scraps are available—is just another form of exploratory sketching. Tim Brown, the chief executive of the design firm IDEO, believes in "building to think." Brown says, "Through the act of making things, we find that we learn about ideas." Design researchers are discovering that the very act of tinkering with materials and objects can be an important part of the learning and discovery process. The design term "thinkering" refers to the ways in which people tend to learn and develop new ideas as they play around with whatever it is they're working on. (The phenomenon is currently being studied by the Illinois Institute of Technology (IIT) at its Institute of Design and, in particular, by professors Dale Fahnstrom and Greg Prygrocki, who are trying to figure out how to create "thinkering spaces" where anyone—including, say, kids in the local library—can learn as they tinker with available objects and materials.)

Learning comes not just from the act of thinkering but also from the feedback that results. Each time Brown's design firm makes a low-cost preliminary model of something and passes it around—particularly among potential customers—the designers invariably learn "that we didn't know as much as we thought we did about the idea," Brown says. "You learn about the failings of your ideas quickly and cheaply."

One of the biggest benefits of making drawings or models may be that it tends to generate forward momentum and progress around an idea, by causing others to respond in some way. "Just by making your thought processes and inventions visible," Mau says, "you automatically affect the people around you—getting them to engage with the idea, inspiring them, challenging them."

A sketch or model of an idea seems to hit deeper than a verbal description of that same idea. "A sketch may seem vague, but it is actually incredibly precise," Mau says. "When people see a concept presented visually, they tend to understand it in a way that goes beyond words."

3.2 DIRECTED ATTENTION: WHY WE DON'T SEE THE GORILLA IN OUR MIDST

There's a good reason why pictures communicate so powerfully. A disproportionate amount of brainpower is dedicated to visual processing, which means that what we see has far more impact than what we're told. In fact, we acquire far more information through vision than all other senses combined.

But even with several billion neurons working to process the visual images before us, we miss a lot, according to Colin Ware, who directs the Data

Visualization Research Lab at the University of New Hampshire. It may seem as if we're taking in the full field of vision that lies before us at any given time, but in fact the eye only sees what is directly in front of it. Through rapid movement (occurring about three times per second), the eye constantly jumps from one point of focus to the next. Meanwhile, something important can be happening practically right before our eyes and we may miss it because we're too busy noticing something else.

This is nicely illustrated in a short film, created by the vision researcher Daniel Simons, in which people dressed in white pass a ball to one another, while others, wearing black, do likewise. Viewers of the film are instructed to count the number of passes by the team in white. In the middle of all this, a person wearing a dark gorilla suit clearly walks into the picture and even waves. Yet most viewers don't notice the gorilla at all because they're keeping an eye on the ball as instructed.

What we can conclude from this, according to Professor Ware, "is that directed attention is everything." The eye sees mainly what it is conditioned or compelled to seek out. In the case of the gorilla film, specific instructions were provided, but there are other, subtler ways that the eye can be "directed" to look at certain things, says Ware. The key is to provide "findability"—visual stimuli that are easy to separate out and process. The right mix of low-level visual properties such as color, form, and motion (just enough, but not so much as to overwhelm our limited processing capabilities, says Ware) can direct attention. So can anything that seems familiar: a face, a shape, a pattern.

As Mau sees it, designers are trying to direct attention by creating a pattern or "signal" that can be tuned in to amid all the visual noise and static. The task becomes harder as the visual landscape grows more cluttered with the images and information that constantly bombard us. The designer must continually adjust the "signal to noise ratio" to have any hope of communicating. And it's only going to get harder: Mau notes that by some estimates there will be a *billionfold* increase in the amount of visual stimuli hitting us over the next twenty-five years.

If that sounds like a case of "information overload," it's actually not. What we're really swamped with is raw *data*. For that data to rise to the level of information—for it to be informative—it must be organized, simplified, clarified, or, in a word, *designed*.

Some of that involves cleaning things up and putting them in their proper place by instituting an "information hierarchy"—basically an attempt to organize visual information based upon what people really need to see and when they need to see it. Michael Bierut once observed that the brilliance of Massimo Vignelli's design of the New York City subway map was that it managed to "deliver the necessary information at the point of decision, not before or after." With so much data coming at us in a scattered way, an important function of design is to separate it into the proper buckets or chunks (a technique known as "chunking") and then to arrange those chunks— giving precedence to the critical stuff, holding back what's less important, getting rid of what's unimportant.

No one understands this better than designers working in the digital space these days. The rise of smartphones has heightened the need to compress and distill visual data so that it can be comprehensible on miniature screens. In adapting various Web sites for the cell phone screen format, designers are compelled to home in on interface functions that are most critical, while dispensing with all other distractions. One might expect this would compromise the experience of visiting Web sites while using an iPhone as opposed to a full-size computer screen, but the streamlining can actually improve the experience by making it faster and easier to navigate a site.

Similarly, in the mad scramble to design all those thousands of new applications for smart phones, the challenge isn't so much about coming up with a big idea as making your idea smaller. "You have to reduce, reduce, reduce," says Erica Sadun, an apps designer and author of *The iPhone Developer's Cookbook*. "You're removing fonts, colors, anything that adds clutter. You have to keep asking, 'What is the core purpose of this app, and what can I trim away?' In many ways, it's app design bonsai."

But organizing, streamlining, and simplifying are only part of any visual designer's challenge.

Good design also has to attract and engage. Says Brian Collins: "The question is, how do you become reductionist without being simplistic?" Indeed, when communicating an idea, keeping it simple is good, but not enough. To achieve the best kind of simplicity, one must add as well as subtract, according to John Maeda, president of RISD and author of *The Laws of Simplicity*. Maeda says one key to designing simplicity is to follow an overriding principle he calls "thoughtful reduction"—or "subtracting the obvious, and adding the meaningful."

OF COURSE, there are infinite ways that one might try to add meaning to the visual communication of an idea. But there is one thing many of the designers interviewed for this book agreed upon, more or less, when asked to articulate what helps a design to attract and engage people. "Every piece of design must have a combination of something that's familiar and something that's surprising," Collins says. "The familiar gets us in the door, and the surprising keeps us engaged."

Going back to how the eye and the brain process visual information, it makes sense that the combination of familiarity and novelty is so engaging. Familiarity stops the roving eyeball because, as the visual expert Ware notes, we're always looking for patterns we recognize. But having recognized the familiar, the eye may be inclined to quickly move on—"unless it is given some kind of a visual puzzle that must be figured out," Ware says.

When Milton Glaser designed his famous I ♥ NY logo (which served to provide a glimmer of hope on behalf of New York during the city's doldrums in the 1970s), he knew that by mixing the heart icon with the letters, it would take people a second or two to figure it out. "The mind loves puzzles," Glaser says. "As a species, we're programmed to solve them. So if you create a small difficulty, the inert mind is moved and guided to action."

Picture puzzles can also provide richer clarification. You look at an image,

Are mammograms important?

Average size of breast tumor at diagnosis in early 1980s when only 13 percent of women were getting regular mammograms.

Average size of breast tumor at diagnosis in late 1990s when 60 percent of women were getting regular mammograms.

figure something out, and, as a result, come away with a deeper understanding and a picture now burned into the brain. The designer and writer Richard Saul Wurman is known for filling his books with simple charts and pictograms that visually explain how things work or how one thing relates to another. One of Wurman's personal favorites is from his book *Understanding Healthcare*: an image consisting of just two dots, of different sizes. In response to the question "Are mammograms important?" Wurman allowed the image to serve as the answer.

The reason visual designs and pictures work so well in this context is that it can be easier to show relationships than to describe them, says Neil Cohn, who teaches visual language at Tufts University. A simple drawing or chart "can make it clear to people that 'this is happening because of that' or 'if we take this action, it will lead to that outcome,'" Cohn says.

In Al Gore's 2006 film *An Inconvenient Truth*, the use of high-impact visual design elements helped to clarify otherwise vague cause-and-effect relationships. In Gore's slide show, each image was designed to simplify the complexities while amplifying the threat. Even the memorable live-action scene of Gore rising in a mechanical cherry picker to reach the top of a global temperatures chart was a clever bit of design juxtaposition that put things in perspective. What was made clearest of all in the film was the possible out-come in store for us all if we don't change course. Forget about "hope made visible"—first, to wake everyone up, *An Inconvenient Truth* had to draw a picture of Armageddon.

3.3 THE *M* THAT DID A HEADSTAND IN HOPES OF SAVING THE WORLD

Having effectively put forth that chilling design image, Gore had to then find a more positive and hopeful one to present to the world. One of the people who took on the challenge of helping to do that was Brian Collins. In his design work for Gore's Alliance for Climate Protection, as well as in a sepa-rate campaign to address water conservation, Collins has explored the ways in which a well-chosen visual image—or even just a well-shaped typeface—can stir people to action.

A hyperkinetic man in his mid forties, Collins has been a design geek as far back as he can recall. As a boy he had an inexplicable interest in typog-raphy and cereal box packaging and may well have been the only kid his age who watched *Lost in Space* and knew that the furniture was designed by George Nelson. "I would try to discuss this stuff with other kids and they just had no idea what I was talking about," he says. Collins understood early on that there was magic, and considerable power, in those details.

His grasp of that put him on a fast track in the business world, and Col-lins rose to become design chief at one of Madison Avenue's biggest ad agencies, Ogilvy & Mather, where he used design-driven approaches to mar-ket Dove soap and Hershey's chocolate and BP (British Petroleum) in new ways (for the latter client, his group created a futuristic eco-educational gas station in Los Angeles). But he eventually left the big midtown office and moved to cozier digs downtown at his start-up firm, which works on busi-ness projects and ecological activism.

As Gore's Alliance, formed in 2007, was trying to establish an identity and rally public support for environmental efforts, it turned to an ad specialist, the

Martin Agency, which brought in Collins. Together they collaborated on solving a puzzle: How do you turn a negative (the doomsday scenarios associated with the global warming crisis) into a positive, so that people will take action instead of feeling helpless? They knew it would be critical to emphasize a spirit of community. But they also knew that people today tend to see themselves as strong individuals. Collins thought these two threads could be connected by a larger story expressed through the design of a new logo.

What he ended up with was stripped down and minimalist: just a circle with the word *we* printed inside. But the eye tends to linger on the *w* because there's something familiar and yet strange about it, which takes a second or two to figure out. That *w* in *we* looks like an upside-down *m*.

It was no accident: Collins had the curvy typeface (dubbed Galaxy Alliance) custom made so that the *w* and the *m* would be upside-down mirror images of one another. During brainstorming with the Martin Agency, the idea had surfaced that joining an environmental movement was about moving from "me" to "we"—that it involved a transformational shift in perspective. At the same time, Collins wanted to plant the thought that this transformation wouldn't require a long, arduous journey—in fact, that *m* managed to become a *w* just by doing a simple headstand. Nor did one have to relinquish a sense of self when joining the movement. As Collins points out, when you just turn that symbol upside down, the *m* is still there. "The *me* is living inside the idea of *we*," he explains.

A conventional ad might have tried to explain this somewhat tricky concept in words that would seek to persuade (if the ad could get people to actually read those words). But Collins and his team wanted us to internalize this message: to see the logo, wonder why it looks the way it does, sort it out for ourselves.

He also wanted the "me to we" idea to seem as if it were gradually bubbling up around us as a phenomenon. To that end, he urged the Gore group to put the logo on reusable tote bags, water bottles, and other eco-friendly objects that get carried around by people. This is something that designers are doing more and more as they infiltrate and begin to change the ad business: They're redesigning the shape and form of the vehicle that carries the

message. It doesn't have to be a page in a magazine or a thirty-second spot on TV. It can be an object you hold in your hand, something you wear or see on the street.

"Any movement has to have its own unique symbols and flags, to let people see it exists," Collins says. And as those symbols are seen in more and more places, "it can make the movement seem bigger and more powerful than it actually is."

In creating the "me to we" logo, Collins pored over lots of existing typefaces before deciding he needed to have one made from scratch. Not only did this typeface have to be able to carry off that tricky *m* to *w* flop, it also had to seem "welcoming and not strident," Collins explains. He pointed out the sweep of the curves in the lettering, the playful sans serif styling that suggested quirkiness. All of it was designed to signal to people that the environmental movement is not a gloomy and judgmental world, but rather a welcoming place where you might even find the company enjoyable.

Would people really get all that out of a typeface? Collins smiles the way someone does when hearing a too-familiar question, then wheels his chair over to a nearby lateral file drawer. He pulls out several sheets of paper and wheels back. He holds up two sheets, each with the word HOSPITAL printed on it—one in a no-nonsense typeface (Helvetica, to be precise), the other in handwritten script. "If you were driving down the road and needed medical help," Collins asks, "which sign would you trust?" Then he held up a second set of papers, in the same two typefaces, but now the two signs said APPLES. "But here you would probably choose the handwritten type because the sign looks fresh, of the moment," he says, "and that's the way you want your apples. The sign being handwritten makes it kind of nice and whimsical. But you don't want a whimsical hospital."

Collins's point is that whether we know it or not, we're always looking for clues in every sign and every symbol. We're in a constant and sometimes frantic hunt for understanding, so that we can figure out where to go and what to expect when we get there. When designers do their jobs well (and ethically), it can help us make better decisions. When designers fail to do so—when the clarity is lacking or the meaning muddled—we can end up confused, lost, and sometimes even sick to our stomachs. Collins learned this lesson from one of his students.

3.4 THE RATIO OF MASS TO INFORMATION (OR "WHO'S BEEN TAKING MY MEDS?")

Collins was teaching a graduate class in design at New York's School of Visual Arts a few years back when one of his students, Deborah Adler, informed him that she'd had a change of heart about her master's thesis. Adler had intended to design a hair care product for people with curly hair like hers. But then she decided that in the post-9/11 world, she should be trying to design something more meaningful. She started looking around for a "serious" challenge. She found it lurking in her grandparents' medicine cabinet.

Adler's grandmother had suddenly become nauseous and ill. It was a mystery at first, until a doctor figured out the source of the trouble: The older woman had accidentally taken her husband's medicine, which was the same drug but in a much stronger dose. Adler was curious about this, so she decided to check out her grandparents' medicine cabinet, and it was an eye-opening experience.

"Everything in there looked the same," Adler remembers. The bottles and labels were all identical; the vital information was printed so small "that even someone in her twenties, like me, could've easily mixed the bottles up," Adler says. "And I thought, 'This must be happening to more people than just my grandmother.'" (She was right about that: Research has shown that more than half of Americans make mistakes when taking their medication.)

Adler began doing her own research by collecting all the standard pharmacy bottles she could get ahold of, then analyzing them. She could have been doing the analysis forty-years earlier—incredibly, in all that time, no one had bothered to redesign prescription bottles, and it showed. As part of her process, Adler did what designers often do—she playacted her way through scenarios in which she might need to take medicine quickly, in a crisis. Adler would hurriedly open the cabinet and then ask herself: *Now what would help me find the right bottle? And what would I still need to know, once I found that bottle?*

The first thing she noticed was that as she went through this exercise, she was constantly turning bottles round and round, trying to read them. "It seems like such an obvious thing, but the shape of the bottles was wrong," Adler says. With the cylindrical bottles used for virtually all prescription medicines, the instruction label couldn't go on flat—it had to curve around the bottle, making it much harder to read. So the problem began with the basic form.

As for color, the problem was that all the bottles were the same clear brown and all the labels white. The worst of it, though, was the information on the labels—it was a design nightmare. The type was too small and was presented in a uniform size. No priority was given to more important information. In fact, the only thing that stood out prominently on most labels was arguably the least important piece of information—the pharmacy's logo!

Adler knew from her design studies that the principle of establishing an "information hierarchy"—which gives prominence to information based upon importance—needed to be applied to medicine bottles. Though, as she says, "It wasn't even a matter of design principles so much as common sense. You just have to think about it: What do people immediately need to know when they're looking at a medicine bottle?" She settled on three things: the name of the drug and what it's for; who it's prescribed to; and how to take it. Adler decided to separate that information from everything else and make it easier to read.

That was just the beginning. She designed a flattened shape for the bottles so the label would be readable in one glance. She placed the name of the drug on the spine of the bottle. She created a separate card with information about side effects, which slipped into a slot on the bottle so that it could be kept with the pills. One of her biggest innovations was the introduction of a color-coding system. "I really was excited when I figured out how to apply color," she says. "At first I thought maybe I'd use colors for different types of medication, but then I realized it made more sense to have a different color for each member of the family—kind of like toothbrushes." The final touch, which went on later, was a new set of warning icons, developed with Milton Glaser (Adler worked in his design shop during and after the master's project). The new icons were much more intuitive. For example, to warn pregnant women, the generic caution sign was replaced with a silhouette of a pregnant woman's body.

OLD   NEW

To Adler, most of these changes were, as she notes, commonsensical, though the veteran designer Collins could see that she was addressing some profound design issues. "When a designer is trying to make complex concepts understandable," Collins says, "you have to deal with ratio of mass to information." In this context, the *mass* refers to all the material and resources that go into producing that bottle and label, while the information is all the useful data you can extract from it. The old bottles were heavy on mass and light on info, Collins says. "And what Deborah does, without necessarily thinking of it in these terms, is that she completely shifts that ratio," he says. "Without changing the bottle's mass all that much, she somehow manages to put in ten times as much information. The result is that the object now has much, much greater utility."

Adler's final challenge was to get her design out into the world. She turned her original sketches into homemade prototypes,—using plexitubing and—of all things—dollhouse parts. Never mind the source of the materials; no one had to know. The point is, "you have to make things look real to get people to take them seriously," she says.

With her idea now in a form she could show, she started by approaching government agencies but was frustrated by the bureaucracy. She decided to go straight to the big pharmacy chains, and started with the one that had a design "rep." At her first meeting with Target representatives, Adler showed her prototypes. By the time she walked out, she had a deal. Target rushed the line to market—it hit the shelves in 2005 and looked almost remarkably like Adler's prototypes, with the exception of a slight reshaping of the flat bottle (with help from industrial designer Klaus Rosburg). Adler's ClearRx prescription system has been a staple of Target pharmacies ever since.

If Adler transformed those bottles, well, they changed her, too. Her grandmother became a minor celebrity, appearing in Target ads. And Adler herself began to think of design differently; she decided to look for other ways she could apply design in the health care industry. One of her current projects involves redesigning the packaging of hospital bandages, so that nurses and patients can better see how to apply the many different types of dressings for different wounds. "There's a huge need and opportunity to bring clarity to the whole area of health care," Adler says. "Whether it's the patient going into a pharmacy or the nurse that has to apply a dressing, things are often made overly complicated. Not enough attention is being paid to the end user, and it really is affecting people's lives."

ADLER IS right about that, although the problem extends well beyond health care. Richard Grefé, executive director of AIGA, the professional association for design in America, notes that confusing visual design plagues most of the communication between the U.S. government and its citizens. Whether it's documents from the Census Bureau, paperwork involving Social Security or Medicare, or the notoriously bad design of voting ballots, "about 90 percent of the communication that goes on between government and citizens is awful and really needs to be redesigned," Grefé says.

Having already taken on the effort of redesigning ballots and voting booths, designers in Grefé's group are also looking into ways to redesign the front page of mortgage contracts. Grefé thinks that clearer contracts, using framing techniques to make the financial risks more apparent, might have averted some of the mortgage default problems of recent years. (Of course, the old contracts employed design framing, as well—but typically did so in a manner that tended to *downplay* those risks.)

Having the right information delivered in the clearest, most logical way not only helps people to make important decisions, it can also help them to avoid mistakes—and potential disasters. In his public lectures, the renowned information design expert Edward Tufte has been moved to tears while showing images of the confusing engineering charts used by people working on the ill-fated *Challenger* spacecraft. Tufte has posited that the tragedy might have been averted if the charts had been clearer and more readable.

Collins is of the belief that we may be able to avoid future disasters, including perhaps global ones, by using design to create more awareness and understanding of the gathering threats. He recently launched a project, separate from the Gore alliance, that represents an attempt to "redesign the future of water," he says. The goal is to change the way people think about water, including the mythologies built around it—that it's everywhere, that it's free, that we can do with it as we wish. The facts, of course, say otherwise: Water shortage is a real and growing concern worldwide. But just bombarding people with those facts, Collins says, is apt to be less effective than somehow showing the problem.

So Collins put out a challenge, as part of the 2009 Aspen Design Challenge, to students around the world: *Try to redesign the way people think about water.* The hope is that one powerful piece of design thinking will emerge that could alter the way people think about the problem by making the issue and its urgency crystal clear.

While discussing the project at his office, Collins was, as usual, sketching the whole time—stick figures, circles, arrows, sometimes circles around the arrows. At one point, he drew a picture of a pool with one person in it, then another pool with lots of people—and the same amount of water.

"This is where we were before," he says, pointing to the first sketch. "And this is where we are now," he says of the second.

Looking at Collins's sketch, one couldn't help thinking, *Who wants to be in that second pool, jammed in there with all those unhappy characters? And by the way, what if Mau's angry stick figure is one of them?*

5000 BCE → 1950

1950 → 2010

3.5 "THE BEST HEADACHE EVER"

Mau is a firm believer in the importance of making change visible and clear to people in advance. In his studio, he places great emphasis on the production of sketches, prototypes, collages, posters, and other means of previewing an idea before it is produced. He also teaches this principle to students, such as at a recent class at the School of the Art Institute of Chicago.

The class was part of an experimental project conceived by Mau, and the eight students involved were tasked with trying to come up with ideas on how to redesign Chicago's aging public transportation system. They'd been working on the project for a few weeks at this point and had marshaled their facts and jotted down ideas. But Mau told them they now had to do something more: begin to make their ideas visible through drawings, three-dimensional displays—anything that could then be placed in front of other people to make the case visually.

"You have to scratch an idea out into the world any way you can," Mau told the students, as he took a black marker to a whiteboard and began to sketch a vision of the train car of the future.

The reason Mau was in Chicago at all—the reason his business had taken a dramatic turn in a new direction, no longer focused on books or other objects but entirely on the concept of designing large-scale transformation in business and society—was because of a breakthrough he'd had a few years prior in Toronto. He witnessed, firsthand, the power of a visual display to stir interest in the possibility of change.

By the time of Mau's Massive Change epiphany, he had already established himself as Canada's preeminent designer, owing to his ability to "direct attention." He knew how to lead a reader through the pages of a book or a visitor through the halls of a museum. He also knew how to direct attention to himself and his rising firm. Having been steeped in the heady content of Zone books, Mau had the aura of the intellectual, as well as the designer's instinct to make things clear. The combination made him notable and quotable. As one of his design collaborators says, "Bruce's intelligence was such that it intimidated people just a little, but not too much."

Mau settled into a comfortable home life in Toronto, marrying and starting a family. His first meeting with his future wife, Bisi Williams, an attractive and engaging woman of Jamaican descent, nicely illustrates this phenomenon of directed attention. A friend of Mau's named Steve brought *his* friend Bisi by the studio on the way to lunch. Mau took one look at Bisi and invited himself along. "I have a very clear memory of that lunch," Mau says, "but in that memory, Steve isn't there."

After a period of sustained growth, Mau's studio business hit a rough patch in the early 2000s. His work on the Panamanian biodiversity museum got bogged down in governmental bureaucratic delays (it's expected to finally open in 2011), and several other projects proved unprofitable. Mau was restless, pursuing lots of ideas that interested him but didn't necessarily pan out for the studio. His business manager at the time, Jim Shedden, was growing frustrated. By 2002, Shedden was urging Mau to focus.

That was when Mau came up with the idea for Massive Change, by way of a classic smart recombination. Mau had recently been asked by the Vancouver Art Gallery to create a museum exhibit on the future of design, and at the same time he was considering launching an experimental student program in design for Toronto's George Brown College. He figured he could connect the two small projects into something bigger.

First, he laid down some ground rules with the art museum: This was not going to be an art show. "I said, 'I'll do a show about the future of design, but only if we agree to take aesthetics off the table,'" Mau says, recalling his negotiation with VAG director Kathleen Bartels. Bartels agreed and Mau set out to work on creating what would be the largest exhibition in the museum's seventy-year history, focusing on design's potential to solve problems and change lives. Or, as Mau puts it, "Instead of being about the world of design, it would be about the design of the world."

Mau's original vision for the show was expressed in—what else?—a sketch, on the back of a napkin. The drawing showed two multiringed circles. In one, the inner ring is labeled DESIGN and the outer rings are BUSINESS, CULTURE, and NATURE. But in the second set of circles, DESIGN is the outer ring, encompassing all the others.

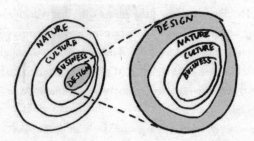

This expressed Mau's growing belief that design was no longer a subset of business, culture, and nature. Instead, because of the exploding capacities—fueled by new technological advancements, increased sharing of knowledge, and good old human ingenuity—it was now possible to design everything from a better corporate structure to better human body parts or a better breed of dog. And so, Mau reasoned, since design could be applied to business, culture, and nature, it was no longer a subset but a determining force; i.e., it had become the big circle. The napkin sketch itself may have been a humble creation, but the idea behind it (which is certainly debatable, especially with regard to design determining nature) was a large and provocative one. "This is probably the most significant diagram of thinking that we've done," Mau says.

To some extent, the Massive Change exhibit was an attempt to elaborate on that napkin sketch and to prove its validity through case studies, statistical data, theories, stories, and a multitude of human creations. Gathering all of that evidence would be a monumental task—but this is where the second part of Mau's plan came into play, involving the college students.

Mau set up an "entrepreneurial learning program," in which the students would tackle a real-world challenge by helping him create the Massive Change exhibit. It was a small class, starting with just eight students the first year and another eight the second, but Mau gave it a grand name: the Institute without Boundaries.

True to form, Mau used some unconventional methods in selecting his IWB team. During interviews, he'd pull out a Scrabble board and engage the interviewee in the game while simultaneously conducting the interview. "It's hard to play Scrabble, be interviewed, and think up bullshit answers all at the same time," Mau explains. It yielded more honest responses, tested multitasking abilities, and also revealed if someone had a sense of humor. "Some people would freak out and say, 'Why are you doing this to me, this isn't how an interview is supposed to go,'" Mau says, mimicking a whine. "But others would just laugh and go with it. If you laughed, you were in."

Mau calculates that the group managed to conduct twenty years' worth of research in a compressed period of less than two years. The team was tasked with sifting through innovations great and small; the goal was to try to chart the ways that design was reinventing transportation, housing, urban life, energy, manufacturing, and biology. No one knew quite where to begin on something so broad and ambitious. In effect, Mau led them all into the deep woods and said, "Now let's try to find our way out."

Greg Van Alstyne, who worked in Mau's studio and helped oversee the project, says that the IWB students almost mounted an insurrection against Mau in the early stages. Mau was well aware of it. "I knew they were angry with me," he says, "and I could understand why. The way most educational programs work is that some teacher tells you what you're supposed to do, and then you do it. That's what the students wanted—someone to tell them what to do." Mau recalls an angry meeting at which he finally raised his voice—he rarely does that—and yelled: "I don't know what to do—and even if I did, I wouldn't tell you!"

THEY DID find their way out of the woods, eventually, and the result was a thing to behold. When the Massive Change exhibit opened in Vancouver in the fall of 2004, *Time* magazine called the show "a cabinet of wonders." Mau filled two floors of the museum with gadgets, gizmos, oversize photo montages, and, in one room, silver balloons. Among the various curiosities on display: a car that could be powered by human pedaling; the LifeStraw water filter, a small device that can be used, like a straw, to sip from polluted waters and drink clean water; an old fuel engine rebuilt as a means of converting biomass into energy; self-healing plastic; Kamen's walking wheelchair; the "bicycle ambulance," designed to bring emergency medical services to areas in sub-Saharan Africa not served by automobile; a regenerated nose, stored in a clear refrigerated case; and a replica of a featherless chicken, bio-designed for warmer climates.

The show was information packed—breaking down and analyzing everything from the global distribution of wealth to shifting urban demographic patterns—but Mau expertly "chunked" it all into digestible nuggets that could be absorbed quickly as you moved through. Each room explored a different subject, and as you entered Mau blasted you with the big idea— usually in the form of a bold declaration such as *We will eradicate poverty*— emblazoned in gigantic Helvetica Neue Bold Condensed type. If you wanted to find out how, you followed the flow of information down through the hierarchy. "We broke a lot of rules about what you're supposed to do in museums," Mau says. "We had eight to ten times as much content as you're supposed to give people."

Using the design technique known as "wayfinding," Mau plotted the path through the overall exhibit so that one idea flowed naturally into the next. He also encouraged forward momentum by planting what he called "devices of wonder" (e.g., the disembodied nose) in plain sight in the next room, so that, he says, "you'd walk into each new place and say, 'What on earth is that?' "

People came out of the show exhausted, exhilarated, or some combination of the two ("It's the best headache I've ever had" was one of the comments shared with the museum curator). Then, too, some came out angry. The response box had messages "from people who basically wanted to kill us by the end of the show," Mau says.

The reviews were also mixed, with both raves and brickbats coming from around the world. Critics dinged it for being arrogant or amateurish; one compared it to "a school project." (Mau was particularly bemused by that comment: "It *was* a school project," he exclaims, laughing). *Time* critic Richard Lacayo, who gave the show a rave review, noted that Mau's optimistic tone, as expressed in those bold proclamations on the walls, could be a bit much. "Every time his wall texts shout, 'We will!' " Lacayo wrote, "a little voice in your head is apt to ask, 'Wanna bet?' "

But if Mau's goal with the show was to make visible a new way of thinking about design with regard to addressing real-world problems, it seemed to work. Over the next few years, a number of major design shows began to pick up and expand upon this theme. The National Design Triennial of 2006 focused on design's connection to real-world issues and social problems. A separate show, Design for the Other 90%—which featured simple design approaches to shelter, water purification, and transportation in the developing world—began touring the United States between 2006 and 2008. Also in 2008, a groundbreaking Museum of Modern Art show, Design and the Elastic Mind, focused on the ways that design was overlapping with science to make technological innovation accessible to the world. These shows were wholly original and very different from Mau's, but the new way of thinking about design that informed them could be seen earlier in Massive Change.

WHAT REALLY surprised Mau was the way Massive Change seemed to resonate beyond the art museum world. Engineers and basement tinkerers saw it as a celebration of the underappreciated art of invention. Environmentalists took note of all the sustainable ideas on display and soon Mau was fielding calls from Chicago mayor Richard Daley, seeking his input on how to make that city the greenest on earth, as well as from a film production team affiliated with the actor Leonardo DiCaprio, inviting Mau to be one of the stars of a film on global warming. There was even a plea for help from a Guatemalan government official. He'd learned of Massive Change and wanted to know what Mau could do for a country going through hard times. (What Mau did was to hop a plane down there and begin setting up a massive exhibition staged in locations throughout the country, highlighting the rich culture and enormous potential of Guatemala. It was *hope made visible* on a national scale.)

Perhaps the strongest response to the show came from the business world, including large companies such as Nokia, MTV, and Coca-Cola. With the show, Mau tapped into something that was just starting to bubble up in business—a sense of urgency with regard to the need to embrace profound change. That change was happening all around these companies, thanks to Internet-empowered consumers, rising environmental concerns, and a marketplace with too many products and not enough innovation. Mau was approached by a number of top executives, all wanting to know more or less the same thing: Could the concept of Massive Change be applied to a major corporation? Was it possible for business to design not just more stuff, but a better and more productive future for itself?

Before long, Mau was taking meetings with these executives, beginning the process of asking them stupid questions and showing them little sketches that just might transform their worlds.

BUSINESS

4. GO DEEP

How do we figure out what people need—before they know they need it?

4.1 GETTING DOWN INTO "THE MUCK AND BEAUTY OF LIFE"

Designers tend to warm to a challenge, so here's one: In business today, nearly two hundred thousand products are introduced each year worldwide. And the vast majority of them are destined to fail. In the United States, for example, the new-product failure rate has been pegged as high as 90 percent.

When products fail, companies scale back or go out of business and people lose their jobs. Store shelves end up cluttered with items no one wants or buys—especially during an economic downturn, when people are selective about their spending. Eventually, those ill-advised creations end up in a landfill somewhere. Is all of this just an unavoidable cost of doing business?

Not necessarily. Designers, being the eternal optimists they are, envision a business environment that operates in a smarter, more efficient manner, thereby producing better results for consumers as well as for companies. This vision of better-designed companies producing better-designed products and services is one that's being embraced more and more by large, established companies, such as Procter & Gamble, as well as by esteemed brands like Nike and Nokia, and small, fast-rising start-ups like the Method soap company.

Talk to the principals at any of these companies and they'll tell you that "design thinking" is playing a key role in the way they operate and make decisions. Does that mean these companies are hiring aesthetically aware designers to dress up the company's offerings in more stylish forms and packages? Absolutely. But the use of design in these companies extends way beyond that surface level. In today's increasingly saturated product marketplace, surface appeal is not enough to elevate me-too products or services above the pack, or to yield sustained success. The harder trick is to use design to get beneath that superficial marketing level and to delve down to core business issues such as: What do people really need? And how can we best provide that?

That business should be focused on offering goods and services people actually want and need may seem obvious. Why would companies do otherwise? The reason is that they often are focused more on their own existing capabilities and objectives than on the needs of others. Heather Fraser, who heads the design program at Toronto's Rotman School of Management, points out that companies logically try to maximize their current capabilities and capacity. "After all, it's where their capital is invested," she notes. Hence, the company that has always made widgets tries to make more of them, in different shapes and colors so that the line can be extended (in this context, designers have been seen as very useful at helping to add those little tweaks—a new style feature here, a fresh package there). In effect, the company is trying to squeeze out as much profit as possible by doing what it has always done, with minor modifications. But in today's dynamic world, those widgets may not be as useful as they once were. They may not be what people really want and need in their lives.

So the issue then becomes: How does a company find out what people truly need? According to corporate conventional wisdom, the only logical thing to do is to ask people. By way of voluminous surveys, focus groups, and various forms of test-marketing, companies inquired: *Do you like what we're doing?* And sometimes: *What should we do differently?* But the problem with the first question was that people didn't necessarily give reliable or useful answers. And the problem with the second question was that people didn't know how to answer at all.

Henry Ford figured out that last part more than a century ago. "If I'd simply asked customers what they wanted," Ford mused, "they would have said, 'a faster horse.'" Of course they'd ask for a horse; they couldn't be expected to ask for something that hadn't been designed or invented yet. It was incumbent upon Ford to design something that people needed, and to do so before they knew they needed it.

As great designers and inventors often do, Ford relied on his instincts to tell him that there was a need for something that didn't exist. At Apple, Steve Jobs and his ace designer, Jonathan Ive, have done likewise in a number of instances. But for businesses in general, relying entirely on the instincts of a lone genius can be too limiting. Never mind the fact that such intuitive wizards are few and far between. What if the genius is great at designing for the needs of certain people but has a blind spot when it comes to the needs of others? Mindful of this, many designers have come to believe that successful design in the business world (and elsewhere, too) is achieved through a marriage of designer's intuition and a deep investigation into people's lives and needs—with emphasis on *deep*.

BRUCE MAU applies the principle of "going deep" in two separate but related contexts. He believes that a designer, in taking on a difficult challenge, must immerse him/herself deep in creative thought on that problem, often for an extended period of time. As noted in Chapter 2, Mau sees this as a kind of journey within the imagination—a time when one may "drift" or "get lost in the woods" while seeking out original ideas and insights that usually can't be accessed with a typical surface-level approach to problem-solving. But Mau also recognizes that designers must, at various points in this process, emerge from creative dreaming and engage with the real world. For that to happen, the designer must again "go deep"—but this time on a very different journey, one that tunnels into the thoughts and ways of other people. "Anthropology and ethnography are essential parts of our work," Mau wrote in his manifesto. "You have to get out there and find people up to their elbows in the muck and beauty of life."

The notion of designers acting as anthropologists or ethnographers is a

The Three Gears of Design

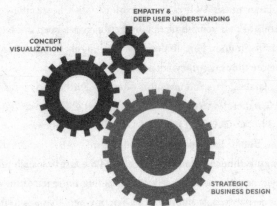

EMPATHY &
DEEP USER UNDERSTANDING

CONCEPT
VISUALIZATION

STRATEGIC
BUSINESS DESIGN

relatively new one, but it has taken hold in a big way, particularly in business. According to the Rotman School's Fraser, user understanding, usually obtained by studying people to try to understand their needs and wants, is the first gear in what Fraser calls the Three Gears of Design for business. In this model, the second gear takes those observations accumulated in that first phase and uses them as the basis for creating sketches and prototypes of new business ideas and innovations. The third gear involves redesigning a company's system of activities in support of gears one and two; it encompasses all the things a company might have to do (add new capabilities, reorganize structurally) in order to realize and optimize the possibilities that emerge from those first two gears of the design process.

When the term "design thinking" is used in business—and it is being used quite a lot these days—it usually refers to some variation on this basic model illustrated by Fraser's Three Gears. Which is to say, it often starts with studying people to try to figure out what they need, proceeds to prototyping as a means of trying to create better products and services, and culminates with a redesign of how the company works. That last gear, not surprisingly, is depicted as the biggest because it can involve transforming a company's overall operations. But the first gear, small as it is in Fraser's model, may actually be the most important. It addresses a question that many companies are grappling with in a time of dynamic technological change, vastly increased business capabilities, and rising competitive pressures. "What I'm hearing from top Fortune 500 executives," says the design and business strategy guru Patrick Whitney, "is that they know how to make just about anything—but they don't know what to make."

4.2 HOW TO CLOSE THE "INNOVATION GAP"

Whitney runs the Institute of Design at the Chicago-based Illinois Institute of Technology, which has emerged as a leading think tank on the evolving relationship between business and design. He says that a growing number of top executives have been coming to his school in the past few years to enroll in design "boot camps" and crash courses. These business chiefs are desperate for design because "they've tried everything else—they've done Six Sigma, they've done TQM," Whitney says, referring to a couple of tried-and-true business methodologies for improving performance (TQM stands for Total Quality Management; Six Sigma is a system of practices aimed at reducing business misfires and defects). The old methods were yielding annual growth rates of 1 or 2 percent a year for these companies— and that was *before* the downturn. Now the issue is not just growth but survival. "And so reluctantly and with trepidation," Whitney says, businesses are coming to terms with the realization that they must find a way to innovate.

And with even more trepidation, they are turning to design as a way to get to innovation—which in turn leads them to places like IIT's Institute of Design, the Rotman School, and the d.school design program at Stanford University. These academic institutions—along with a few individuals, such as the journalist Bruce Nussbaum of *BusinessWeek*—have been at the forefront in elucidating the link between design and innovation.

Innovation is not synonymous with design, but it is closely aligned with it, at least with regard to the way the two terms are used in today's business world. Larry Keeley of the Doblin consulting group, who's also affiliated with the Institute of Design, explains the relationship between the two as follows: Innovation, he says, is a business process that makes a science out of discovery. To do that, the process must harness design capabilities—including the ability to discern what people need and to then give form to those findings.

Keeley and others take great pains to separate the two terms, in part because the words tend to evoke sharply different reactions in the business world. It has been observed that business executives are much more comfortable talking about innovation than about design. *BusinessWeek*'s Nussbaum thinks this is because innovation seems more like a science or an engineering term and is perceived to be "more masculine." Design, meanwhile, is linked to art and femininity; as Nussbaum puts it, design is "associated with drapes."

Innovation may be immensely popular as a business buzzword—Michael Bierut declared that among corporate executives "innovation is the new black"—but the reality is that people talk about it far more than they achieve it. Business people may tend to think of some small tweak in their production process, or a new button added to a gadget, as an innovation. But as the designer Greg Van Alstyne notes, an innovation, by definition, should produce significant change in the marketplace and/or in people's lives.

By that standard, new technology doesn't necessarily result in innovation unless it is utilized in a way that brings about meaningful change. That distinction has been lost on many in the business world, says Whitney. "A lot of business people figured that innovation and technology were the same thing—that if you had the capability of making more stuff, or putting in more features, you were innovating."

In actuality, however, those expanding business and tech capabilities—enabling companies to easily produce more "stuff"—have exacerbated a problem that Whitney refers to as "the innovation gap." As companies began to create an ever-expanding array of offerings, they were simultaneously becoming more and more out of touch with consumer wants and needs.

Interestingly, Whitney's theory suggests that a growth in one area of knowledge directly correlates to a decline in the other. Ergo, the "smarter" that companies became about using technology and efficient business processes to make things, the "dumber" they became about the people on the receiving end of all that output. "As companies gave people more and more products to choose from, they made those people's lives more complicated and harder to keep track of," Whitney says. "We know less about how a family lives now than in the 1950s, because people are living with so much more complexity and have so many more choices." And at the same time, by giving consumers so much more to choose from, producers shifted the balance of power in favor of those consumers—who became ever more particular and demanding in terms of what they wanted. All of this, Whitney says, has continued to widen the innovation gap—that chasm between the organizational knowledge of how to make things and the knowledge of what people want.

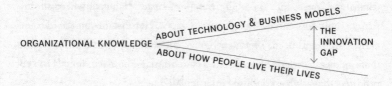

ORGANIZATIONAL KNOWLEDGE   ABOUT TECHNOLOGY & BUSINESS MODELS
ABOUT HOW PEOPLE LIVE THEIR LIVES
THE INNOVATION GAP

DESIGN THINKING—and in particular that first gear, focusing on consumer understanding—is the key to closing the innovation gap, Whitney maintains. If a company can gain a better understanding of how people are actually living today, and get a sense of what may be missing from their lives, the company can then attempt to bring its organizational capabilities and knowledge to bear on that particular need instead of flailing around with product introductions that aren't needed.

Consider as an example the highly successful product introduction of the Nike Plus system, which connected Nike running shoes with Apple iPod technology, enabling runners to keep tabs on calories burned and performance times (through a sensor embedded in the shoe that transmits to the iPod) while also listening to whatever it is that pumps them up. This wasn't just another product line extension. It was an innovation, because it changed the lives of a core group of Nike customers. How did Nike know it was needed? Because the company employs design researchers, who go out into the field (and onto the racetracks, and the asphalt courts) to observe behavior. The researchers learned that runners were starved for ways to measure their own performance—and also noticed that most of them wore iPods. Observation led to the concept, which (after no small amount of design work and haggling with Apple) led to the innovation.

Nike has achieved a number of innovations through this type of in-the-trenches design research. Kathleen Brandenburg, a designer with the firm IA Collaborative who has worked with Nike on a number of projects, goes wherever the athletes go, bringing her notepad or, in some cases, a video camera. Brandenburg's immersive research for clients such as Nike, Gatorade, and others has dropped her into the midst of pickup basketball games and, once, into the chilly waters of Lake Michigan, where she researched outdoor athletes kayaking at five AM. Her insights on the idiosyncratic habits and needs of athletes—e.g., the need for special wristwatches that can measure performance six ways to Sunday—have led directly to new products that addressed those "unarticulated needs."

And therein lies gold for designers and the companies that work with them: If you can uncover and then satisfy unarticulated needs, the resulting products and services will tend to be innovative, distinctive, and coveted. And if you can do all of this somewhat consistently, Whitney notes, you begin to close the innovation gap, bringing the company's capabilities and efforts more in line with what's needed in people's lives. And—oh, yes—there is money to be made in all of this, too.

At least, that is what the data suggests. In the past, it was unclear whether design really could be tied to profitability, but a number of more recent studies have begun to examine the success rate of companies that are design driven (meaning they place a strong emphasis on designing a better product and overall experience for customers), as compared to those that are not. According to the American research firm Peer Insight, design-driven companies outperformed the Standard & Poor's 500 by a ten-to-one margin. In the United Kingdom, a study by the Design Council found that design-driven British companies outperformed competitors by 200 percent.

Such studies notwithstanding, it is still difficult to directly measure the value of design within a company's overall performance. But the track record of design-driven companies suggests that there are compelling reasons for companies to implement the use of design research and prototyping. It does, however, require a radical shift in perspective and approach.

The design-driven model turns the conventional business model upside down in some ways: Companies accustomed to producing what they've always made and then using marketing to try to foist it on the world must shift to more of a listening-learning-adapting mode. And they must be more willing to wade into that "muck" of everyday life—or, at least, to have intrepid design researchers do that for them.

4.3 THE MYSTERY OF THE MOWED TOES AND THE RISE OF THE DESIGN MIND

Designers have always, to some extent, relied on empathy when trying to create objects and products that could serve the needs of others. But classic designers such as Charles Eames or Eliot Noyes didn't employ today's more formalized and elaborate observational research techniques; they tended to rely on their own keen instincts (which served them well).

It wasn't until the second half of the twentieth century that design research—a social science stew blending ergonomics, psychology, and ethnography—came to the fore. The seeds were planted in World War II. At the time, the cutting-edge military equipment then being produced—tanks, submarines, aircraft—was high on technological sophistication but low on usability. Pilots, in particular, were having trouble figuring out how to operate the faster, fuller-featured planes. Sometimes the issue was not complexity but comfort—soldiers had to spend long periods of time inside tanks and subs, so these enclosed spaces needed to be designed with this in mind. Top designers such as Henry Dreyfuss were brought on board to assist with ergonomics issues, while at the same time cognitive specialists and psychologists helped to take on the challenge of trying to design control panels and mechanisms so that they'd make sense to the soldiers using them. The design principle of "mapping"—which focuses on the intuitive relationship between controls and their effects—was critical: A pilot needed to understand, without having to figure it out, that by moving a lever *this* way the wing would tip *that* way.

These design lessons learned during the war—that an awareness of "human factors" could help make advanced technology more acceptable and usable for ordinary people—were reinforced in the decades that followed. The burgeoning high-tech industry relied on increasingly sophisticated design research to organize and simplify the interaction between human beings and computers. Beyond the business world, governments began to explore the possibilities for using new forms of design research to solve a variety of idiosyncratic problems. For example, in the early 1980s, the British government observed a high incidence of people cutting off their own toes with their lawn mowers. Why was it happening? What could be done about it? A psychologist named Jane Fulton Suri was tasked with answering those questions.

FULTON SURI would eventually become a preeminent figure in the realm of design research, but at the time of the "missing toes" incidents, she was not involved in designing anything. In her work with government agencies, her role was to serve as a kind of detective sent to investigate various types of accidents to try to understand why they happened and how they might be prevented.

Fulton Suri did not rely simply on accident reports. Her approach was to go right to the source, interviewing accident victims and asking them, ever so politely, to re-create the mishap. "I wanted to understand the context in which these problems occurred," she says. "That means you have to get out there and go to people's homes, and then get them off the sofa. I'd say, 'Okay, now take me into the toolshed. Show me where you were and what you were doing.'"

Fulton Suri utilized the full range of her psychology skills: "Perception, cognition, interrelating with people and getting them to open up and remember—I was using all of that," she says. She had to put people at ease "so that they felt free to share the truth of what happened." And she was very careful to avoid judging them. "For many of these people, it was kind of cathartic in a way because they had someone who was interested in them but who wasn't going to tell them how stupid they'd been. I would say, 'Right, now let's figure this out together so it doesn't happen again.'"

When Fulton Suri began looking into the series of incidents in which people were losing toes to the rotary blade of a popular type of lightweight electric lawn mower, the first thing she observed was that the compact upright machine resembled a vacuum cleaner in some ways. "This led to a sense among people that it was a casual piece of equipment like a Hoover," she says. As she had accident victims reenact their experiences with the machine, she found they were using it like a Hoover—running it back and forth in front of where their feet were planted (and to make matters worse, people were using it "while wearing their carpet slippers or flip-flops," she says). During these reenacted scenarios, Fulton Suri also learned that a number of accidents happened not during mowing but while people were carrying the machine across the yard. The mower had been designed to start up when you gripped the handle, "so people were constantly turning it on accidentally," sometimes while the blades were near their feet.

These findings led to the establishment of new government standards requiring built-in safety features on products such as the electric mower; in subsequent versions, users had to perform a much more deliberate "double action" (pressing a button and holding a lever simultaneously) to turn on the machine. The wave of toe-mowing incidents came to an end.

Fulton Suri took a similar approach with incidents involving handsaws and severed fingers, as well as traffic accidents in which motorcycles were hit by cars (directed attention was to blame: drivers changing lanes were so focused on the big truck in their side view mirror that they would fail to see the motorcycle in front of the truck). Fulton Suri's work led to government mandates for daytime running lights on motorcycles.

While Fulton Suri took pride in her work at the time, there was one aspect of it that left her feeling dissatisfied. Both in her work for the government and in earlier work she'd done consulting with companies on product safety issues, "I was being brought into the process *after* something had already been designed," she says. "And it just always seemed obvious to me that if you could have seen these problems in the earlier design phases, many of these problems could have been avoided."

In the late 1980s, while Fulton Suri was in the United States, she met the designer Bill Moggridge and then later was introduced to another top American designer, David Kelley, known for, among other things, developing the first computer mouse for Apple. Moggridge and Kelley both had their own design firms but would eventually join forces to start a new one named IDEO in 1991. A few years prior to the merger, Moggridge hired Fulton Suri to help his firm, ID Two, apply more psychology-based research to its design projects.

"For a design firm to have a psychologist on staff was unusual at the time and it was a risk," Fulton Suri says today. "Most of my work had been about looking at things once they were proven to be problematic. There was no tradition of having someone from the social sciences come in at the early stages of design, to try to anticipate problems and needs in advance."

When IDEO was formed, Fulton Suri came aboard and helped the fast-rising firm to break new ground in using empathic research at the up-front stages of design, as well as throughout prototyping and refinement. IDEO eventually began to refer to this research-driven method as "design thinking." The firm's co-chief, Kelley, didn't need to be sold on this approach—he'd come out of the design education program at Stanford University, which also stressed the importance of human factors in its design teachings.

At IDEO, Kelley was excited by the potential of taking the Stanford need-finding approach a step further with the aid of trained social scientists. "It was an important change," he says, "because it meant that as a designer, instead of feeling that you had to figure everything out from your own point of view, you could actually go looking for latent needs by studying people."

And so much the better if you could do that accompanied by someone with a keen eye for spotting those needs. During one of his early encounters with Fulton Suri, Kelley relates, "I was standing next to her and we happened to be in front of a Pepsi machine, and Jane said to me, 'Do you think it's right that the Pepsi can comes out at your ankles?' And I thought, *Not once did it ever occur to me it's weird that people have to kneel down to get their drink*. At that moment I realized that working with Jane was going to really help take off my blinders."

4.4 "A BLINDING GLIMPSE OF THE BLEEDING OBVIOUS"

IDEO began to reinvent the whole process of studying consumers, moving away from classic, well-established business research tools such as the focus group. Although group interviews and other forms of surveys were immensely popular with many businesses, IDEO's designers shared the old Henry Ford view that it does no good to ask people what they want or need, because oftentimes they just don't know. IDEO also maintained that focus group participants were often less than candid. According to Tom Kelley (David's brother and a partner in the firm), people were apt to hold back from being truthful when evaluating ideas or products "for the same reasons that dinner guests won't tell you your meat loaf tastes like sawdust." Politeness, peer pressure within the group, and fear of saying the wrong thing all tended to influence responses and compromise the research value of focus groups.

Fulton Suri and others at IDEO felt it was often more important just to watch people, without necessarily asking questions; the preference was for "observation over inquiry." And instead of bringing subjects into a sterile interview room or office, the design researchers wanted to observe people in a more natural habitat—which meant they would effectively move into people's homes (with permission, of course), eat meals with the family, follow people around the kitchen, and sometimes shadow them on daily rounds of shopping, work, or errands. These extensive research sessions, which could go on for days or weeks, became known as "deep dives." When IDEO did convene focus groups, the design researchers turned them into "unfocus groups." Instead of just asking questions, participants would be encouraged to tell stories, draw pictures, assemble collages—and all the while, little snippets of side conversations between group members were recorded, to get a sense of what they were *really* thinking.

The researchers didn't just watch and listen; they also tried to immerse themselves in the experiences of the people they were studying. For example, when IDEO began working on hospital redesigns, the firm's designers spent time flat on their backs in hospital beds in order to see things from the patient viewpoint. (What they saw mostly was the ceiling, of course—which led them to recommend to their hospital clients that the ceilings be decorated and used to post information for patients.) Similarly, when asked to help in designing furniture for young children, one IDEO designer followed kids though their play days and, upon observing that they liked to huddle under tables, got down under the table himself—whereupon he came up with a bestselling idea for a storage device that attached under tables and allowed kids to stash toys there.

In each of these instances, the designers were trying to see things through the eyes of the people they were designing for, notes Paul Bennett, a creative director in the London office IDEO opened as it expanded beyond its Silicon Valley base. Just by taking a different angle or perspective, Bennett says, the designers hoped they could catch what Bennett called "a blinding glimpse of the bleeding obvious."

WHILE IDEO was ahead of the pack when it came to expanding and formalizing design research, the firm wasn't the only one pushing in the direction of more user-centered design. The philosophy had taken hold at European companies such as Philips and was also being embraced by start-up design shops such as Smart Design in New York. Smart Design was rooted in the notion that product design should respond to the sometimes-mundane but nevertheless important needs of all kinds of people.

The firm's cofounder, Davin Stowell, designed his first hit product—it was anything but glamorous—based on his own experience as a college student who ate meals on the fly and hated cleaning up dishes. Stowell conceived a cooking pot that doubled as a bowl so that you could eat out of it—"one less dish to wash," Stowell reasoned. When he showed the idea to Corning, where he was interning at the time, they in turn subjected it to a focus group, where it was dismissed as being "uncivilized." Corning eventually took a flier on it anyway, and it ended up being one of their bestselling products.

Once Smart was up and running in the 1980s, Stowell continued to base his design work on close observation of people as they went about everyday tasks. The company was using empathic research and ethnography "but we didn't have a name for it," says Stowell, a bearded, soft-spoken and slightly fidgety man. "All we knew was, we were going into people's homes to look at how they used things. You watch long enough, you learn little things that people never think to say—'This bowl makes a funny noise when it scrapes against the counter.' That kind of thing."

Stowell's firm was hired in 1990 to create the first OXO Good Grips product—specifically, the potato peeler mentioned in Chapter 1, which was being funded by an entrepreneur whose wife's arthritis made it difficult for her to use an ordinary peeler. Stowell and his design partners had to immerse themselves in the experience of the people they were designing for, so they went to the local arthritis foundation and "hung around the library trying to meet people," recalls Dan Formosa, one of Smart's cofounders. The designers logged many hours talking to arthritics and watching them work with their hands.

As they moved from observation into the "second gear of design" on the OXO peeler—prototyping—Smart created hundreds of mock-ups of the peeler before getting it right (which is actually not so bad when you consider that the designer James Dyson made *five thousand* prototypes before figuring out his revolutionary vacuum cleaner). Stowell's firm didn't have the kind of high-tech 3-D printing machines it now has for making prototypes. Foam replicas of the peeler were mostly carved by hand, then taken back out into the real world to see what people thought.

As the peeler's handle took shape, the designers would continue to add a little carved ridge here and a groove there, trying to mold a better grip. (Designers have been obsessing over this sort of thing for at least thirteen thousand years, apparently: When a cache of buried Stone Age hand tools was unearthed recently in a backyard in Colorado, the homeowner who found the tools was struck by the way "they fit perfectly in your palm, and your fingers curl over just where they should.") Eventually, the OXO designers ended up carving in a full set of squeezable fins on each side of the handle. They also discovered that an oval-shaped handle, as opposed to a round one, did not roll in the palm of the hand as much, which proved helpful to people with arthritic fingers.

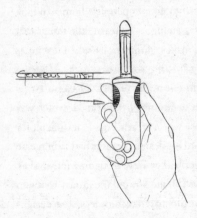

When they'd finished designing the peeler, Stowell knew they'd done something right, because "as soon as you looked at it, you just wanted to squeeze it." OXO had no ad budget to speak of, but cleverly promoted the product by setting up demos in stores, with sample peelers placed next to bowls full of carrots. By word of mouth the product became an immediate sensation. Funny thing: It turned out that *everybody* liked a kitchen tool that sat comfortably in the hand. Smart and OXO stumbled upon a design truth that is now codified as principle: If you can satisfy the needs of even the most challenging end users—sometimes known as the "extreme users"—you can achieve a form of universal design. And if your product is universally loved, there is a good chance you will become universally wealthy.

4.5 WATCHING THE TRANSVESTITES. AND DISCOVERING THE "BUTT BRUSH."

You didn't have to be arthritic to be an "extreme user." In its research, IDEO was discovering that people who happened to be at either end of the spectrum when it came to a particular activity or product—either they were hardcore fanatics or completely inept—proved to be the best to study. The fanatics tended to find new and better ways to use existing products, and designers could learn from that. The inept subjects made mistakes and used tools the wrong way, which also proved instructive. IDEO's chief executive Tim Brown recalls that when IDEO was designing a pizza cutter intended mainly for adults, the firm brought together small children and watched them try to cut pizza. The kids had trouble getting the right angle or leverage, "which showed us, in an exaggerated way, the same kinds of problems adults were likely to have."

In seeking out the extremes, IDEO found itself in the company of quirky characters. For dental hygiene products, the firm was apt to study the person who has "seven kinds of toothpaste" alongside someone who is toothless. For cosmetics, it was interested in women who shunned makeup altogether as well as those who slathered it on (and the firm discovered that when it came to finding fresh and creative uses of makeup, you couldn't do better than watching a transvestite).

The extreme users weren't the only ones worth watching, though. In general, IDEO found that people tend to be very creative at coming up with homemade solutions to fill the gaps not being addressed by the products and services in their lives. People did this through "work-arounds"—if a product didn't work as well as it might, people would figure out how to somehow make it better, using elbow grease, ingenuity, and, if necessary, duct tape.

An interesting example of this (which didn't involve IDEO) occurred when the company Rubbermaid was developing a new line of walkers for senior citizens. Rubbermaid's design researchers went into nursing homes to ask how the walkers might be improved and were told that the walkers were already fine and needed no improvement. But as the researchers looked around, they observed that a woman had tied a bicycle basket to her walker with shoestrings; a man had made a cell phone holder and attached it with duct tape; another had a cup holder from a car jimmied on. Rubbermaid got the message, and designed its new line of walkers with a flexible mesh pouch attached.

AS IDEO took its design research findings to corporate clients, it sometimes served as an eye-opener for companies who'd been designing and producing products the wrong way without ever knowing it. For example, everyone had

assumed children's toothbrushes should have small handles to fit their small hands. But as IDEO actually watched kids brush their teeth (no one ever said design research was a thrill ride) they noticed the toothbrush handle rolling around in small palms and—click—a light switched on: Small kids actually needed *thicker* toothbrush handles that require less of a closed fist. The thick-handled toothbrush became an innovative breakthrough product for Oral-B.

Some of IDEO's methods and principles began to filter out into the larger business world and were embraced by top companies such as Procter & Gamble. About ten years ago, P&G started working closely with IDEO, Continuum, and other design consultants to implement a full design research program within the company. It changed the way P&G began to look at its customers and their needs, which allowed the company to spot a number of opportunities to fill unarticulated needs. The company learned through observational research that people were having a hard time cleaning difficult-to-reach areas of bathrooms, so it came out with a new kind of flexible mop that could do the job. P&G also learned that busy people were looking for ways to freshen their clothes on the fly. It responded by creating

Tide to Go in a stick form, and later with a full line of "Swash" sprays that could clean and freshen clothes without washing them.

With ethnographers and designers spending increasing amounts of time watching people in all kinds of situations—at home, at work, getting medical treatment, boarding planes, checking into hotels, shopping in stores—the research invariably revealed odd surprises, relating to quirks of human behavior that had previously gone undetected. The phenomenon of the "butt brush" effect was one such example. Paco Underhill, a specialist in observing people's behavior in stores, discovered that if a shopper paused to consider buying an item on the shelf and was lightly brushed from behind by someone passing down the aisle, this minor disruption had a major effect—more often than not, the original shopper would immediately put down whatever he/she was looking at and abandon that location. Underhill's observation led to store redesigns that incorporate wider aisles.

Design research has also begun to help companies get a better understanding of some of the surprising habits and customs of the world community. Nokia, in particular, has undertaken an aggressive global ethnography effort. The Finnish cell phone giant has established design research outposts in emerging markets such as China, India, and Indonesia. Its field studies in Uganda revealed that the cell phone was a shared tool—with one phone passed among family and friends. Nokia responded with phones designed for heavy handling and with built-in shared address books. In China, Nokia's researchers shadowed users in dark apartment hallways—and saw that people were using the cell phone screen to illuminate their way down the hall. The next generation of phones for China had a one-touch flashlight feature.

4.6 VOYEURS, POSEURS, AND STALKERS—OH MY!

Nokia is so good at shadowing people and insinuating itself into their lives that one of the company's lead design researchers, Jan Chipchase, has referred to himself as "an authorized stalker." He's joking about that, but some people do feel that the new wave of home invasions by designers is over-done—and may even border on creepy.

"I think a lot of this ethnography is being done by designers because they have fun doing it—they get to be voyeurs," the design critic Donald Norman says. While Norman believes that designers should get out of their studios and interact with people, he doesn't think that necessarily requires hanging out in people's bedrooms. (And ethnography can, indeed, occasionally go to those titillating extremes: *Advertising Age* reported on an example in which a small design research firm did a deep dive on behalf of the company Unilever for one of its male hygiene products. The ethnographers presented "explicit insights into the sex lives and psyches of young males," and as part of the research presentation Unilever executives were invited to role-play by sharing "stories of losing their virginity." A Unilever executive told *Ad Age* she gets ongoing requests for the DVD of that memorable meeting.)

As demand for design research has grown, it has given rise to, in the words of one designer, ethnographic "poseurs" who tend to lack the train-ing and rigor needed to do the job properly. When not approached profes-sionally and objectively, design research can become an exercise in rationalization—wherein the findings are spun in order to yield whatever a particular client may wish to hear.

Along these lines, there is some suspicion in design circles that IDEO's emphasis on the role of research may be "designed to sell design," because conservative clients tend to feel more comfortable when the art of design is bolstered by complex terminology, studies, jargon, and process. "It's a lot of hooey," the Eliot Noyes biographer and designer Gordon Bruce says of the new design research. He sees it as a crutch for designers who lack the stature, confidence, and unshakable faith in creativity that Noyes, Eames, and the rest of the old guard had.

Some of today's star designers, such as Yves Behar, are wary of ethnog-raphy mania because it can seem to put talent and creative instinct in a sec-ondary position. No matter how much one may try to inject research methodology into the field, Behar maintains, "good design comes from people, not processes."

MAU SEEMS to be of two minds on all this. On the one hand, he believes that designers must get out of the studio and immerse themselves in the world they would deign to improve, and he does so himself. He went to live among the Guatemalans while doing design work for that country and lost himself in Panamanian nature preserves while working on that country's biodiversity museum. He believes that, in the basic equation of design, watching and listening to people is the input that ultimately leads to the output.

On the other hand, Mau has supreme confidence in his own instincts. "I know this is going to sound arrogant," he says, "but often, the first idea that I come up with—the one that surfaces in the fog of war—is the best one." So Mau's working model continues to put design intuition first: The idea is to start out thinking of original ideas and solutions, then take those ideas out into the field to see if they match up with what people are thinking and saying, while always remaining open to new ideas that might emerge in the research and cause you to change direction. Mau sketched his model, showing that instead of starting with research (*top drawing*), he starts with design speculation and then uses research at intervals throughout the process.

For example, when Mau was approached by Arizona State University

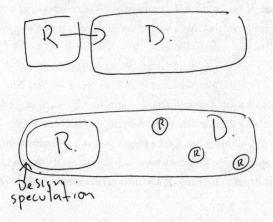

and its dynamic new president, Michael Crow, who was trying to transform ASU from a renowned "party school" into a center of entrepreneurial learning and cutting-edge research, Crow told Mau he was looking for a bold way to signal to the world that the school was undertaking its own form of massive change. Mau flashed on an idea right away: What if a university decided to marshal its academic resources and student energies toward taking on "one hundred challenges" facing today's world? Mau thought it could provide a mission that the whole school, and its fund-raisers, could rally behind. The program could be announced to the world by way of a dramatic event not unlike Mau's Massive Change show.

While Mau was contemplating this idea, his firm's research team began to do a deep dive at the university, spending time with professors and students and using an IDEO method called "cocktail parties," which brought people together in casual settings for relaxed conversations. Over the course of the research, a half dozen different concepts emerged about how ASU might reinvent itself. All were presented for feedback from people at ASU, but after an exhaustive winnowing process, everything kept coming back to the hundred challenges; Mau's original idea prevailed.

This doesn't mean the research wasn't useful, Mau says; his idea needed to be tested against other ideas and validated. And the original concept gained some depth and detail along the way, giving the idea some extra fuel. "But the fuel—which you get from research and listening to people—is never enough by itself," Mau maintains. "There has to be a spark that ignites everything, and that's where the designer's talent and vision come into play."

WHETHER DESIGN research should happen up front, as in the IDEO and Three Gears models, or later in the process, as with Mau, is open to debate. What matters is that it becomes a bigger part of the business model at the many companies that have thus far remained too cut off from what customers want and need. Dev Patnaik, a partner at the design firm Jump Associates

BRUCE MAU

"Design Is the Art of Science"

I believe in science and art, and the promise and potential of design to bring them together to change the world.

In our intellectual institutions, and our society in general, science and art live mostly separate lives—developing separate worldviews, distinct methods, specialist language, and segregated communities of thought and practice.

But the more I work as a designer—a practice that demands the constant negotiation of the boundaries and intersections of these two worlds, where every creative outcome relies on a scientific base—the more deeply committed I am to the foundation of science.

The historic scientific project has built a common language of human knowledge that allows any scientist working anywhere to challenge and revise and add to our shared resource. It was open source before that idea was named. And it is on this human foundation of knowledge that we solve the challenges we face as a global society.

It is science that has opened up the electromagnetic spectrum so that we can explore the universe in all of its complexity. It is science that has allowed us the wonders of communication so that you can read my words, and we can connect across borders and cultures and time zones. And that has built the tools to overcome diseases like smallpox and SARS, and soon polio and others. And that has also built the pathway out of the dark night of ignorance, mythology, and super-stition, allowing us to replace imaginary stories with hard-won empirical knowledge.

But my commitment to scientific knowledge in no way diminishes my belief in the mystery and power of the arts.

It is art that sings to us and opens our hearts to one another. It is art that gives meaning to things that would otherwise go unnoticed. And that connects us to our past. And that laughs at our hubris and limitations, while speaking to us of the darkness we cannot say out loud.

In the end, it is art that allows us to understand, express, and share science. While science works to order the *matter* of the world, art orders the *meaning* of the world.

In my practice of design these two worlds of meaning and matter, of aesthetics and scientific knowledge, of quality and quantity, of mystery and certainty, of intuition and expertise, come together to create new possibilities for shaping our world.

As a designer, everything I do draws on the arts and the aesthetic dimension of cultural life, and also rests on the foundation of the scientific project. Everything I do summons up the mystery of beauty and the history of form, but demands the technical base of knowledge for its success. Nothing we make can succeed that does not draw on our technical expertise and the science of material and energy and process. Nothing we make is relevant if it doesn't "work." But nothing we make is relevant if it is not also cultural, and beautiful in some dimension.

Buckminster Fuller once wrote, "When I am working on a problem, I never think about beauty, but when I have finished, if the solution is not beautiful, I know it is wrong."

Without the dimension that we call art—color and texture, form and material, juxtaposition and composition, humor and metaphor—the full potential is somehow unrealized. No matter how efficient or effective, no matter how "smart," without the language of art, the things we make are limited in meaning and potential. It is art that connects to our life, to human needs and emotions—and that allows us to build a bridge to new possibilities.

At no time in human history has the potential for designing solutions that contribute to the benefit of humankind been greater than it is today. Because of the growing body of knowledge in both science and the arts, our possibilities will be even greater tomorrow.

and author of the book *Wired to Care*, believes many businesses have suffered because they simply can't relate to the world outside the corporation. At the big three American automakers, for example, there's little sense of what it's like to shop for a car or maintain one for long periods of time because, Patnaik notes, top employees have had company-made cars regularly given to them or sold to them at a discount. A similar kind of insularity can be found at all kinds of companies, according to Patnaik: "You have folks marketing tomato sauce who only dine at fine Italian restaurants and have not opened a jar of spaghetti sauce in their own house in years."

To begin to overcome this problem, companies don't necessarily have to dive into the deep end of design research, but they must at least venture out into the world a bit more. "It's important to get outside your office and visit the places where real people live their lives," Patnaik says. "Spend time with those people and talk about what *they're* interested in." As IDEO's Jane Fulton Suri points out, it is a somewhat unfamiliar activity for business people to spend time interacting with outside people, listening to them and observing their behavior: "It may not feel like legitimate work," Fulton Suri wrote in her book, *Thoughtless Acts*. People in the business world must get used to the idea that "it's okay to observe," she says.

To unlock the business potential of observational research, it's critical to be able to bring outside learning back into the company and share it. Patnaik believes companies should plaster the walls with pictures of customers, artifacts of how they live, stories that have been shared; you should "create an office that looks like a shrine to the people the company serves," he says. (OXO Good Grips does this by having employees collect lost gloves from around the world, which are then displayed on a wall at the company's New York headquarters—to serve as a constant visual reminder of all the different hands that the company's products need to accommodate.)

Companies can incentivize empathic research by rewarding employees based on how much time they spend outside the office with customers; they can also make sure managers stay in touch with customers by having them take rotations on frontline jobs. But Patnaik says it may be up to the individual employee to take the initiative when it comes to bringing more empathy to the company. "We've seen lots of cases where a small group or even an individual shifted the direction of an organization by modeling the behavior they wanted to see everyone adopt," he says.

The benefits of bringing this approach to a company can extend beyond the bottom line. As businesses spend more time among people of varying types, products are more likely to be geared to those who were previously ignored or ill-served by the business world. OXO's empathic-driven approach to making products "for the rest of us" has been embraced by a growing number of companies that are using design research to better cater to the needs of specific communities such as women, minorities, and senior citizens. Brandenburg recently worked with Hewlett-Packard on a project that focused on how to connect the elderly with family and friends by offering tech products designed to be simpler and more accessible to seniors. She observed all the small things that can cause older people to be intimidated by technology—the placements of the controls, the overabundance of features—and reported back to the client, which promptly began work on a simpler product geared to that market segment.

Of course, design research also lends itself to exploitive purposes: Companies can use ethnography to identify ways to push various junk offerings. "There is definitely an ethical issue involved," Jane Fulton Suri says. "It's not as simple as saying that if we only know what people want we'll give them better products, because we may just end up giving them high-fat foods and sugary drinks. But I think we should be looking for a deeper truth in people's lives—because deep down, people want to be healthy and happy and to live a good life. If we can figure out how to help them do that, then we're fulfilling the potential of design."

Perhaps it comes down to distinguishing between *want* and *need*. Advertisers have historically tended to focus on what we want, or think we want, or what they've convinced us to want. Designers, at least in their better moments, have paid a bit more attention to what we actually need.

So if the "deep dives" into our lives are guided more by designers than by marketing executives, we may gradually see a reversal of an existing business process that, as IDEO's Brown says, "has been focused on inventing stuff and then saying, 'Now, who can we sell this to?'" In the new model, business goes out to the people before making anything. And instead of persuading those people about what they want, business shuts up, takes a seat in the corner, and starts paying attention.

5. WORK THE METAPHOR

Realizing what a brand or business is really
about—then bringing it to life through
designed experiences

5.1 CHOCOLATE COINS AND LEMONADE STANDS

In the Pacific Northwest of the United States, there is a regional chain of
banks bearing the unusual name Umpqua. The bank offers CDs, but not
necessarily the kind you'd expect: Right out in the middle of the bank floor,
Umpqua sells music discs of local rock bands (the bank also helps sponsor
those bands). Depending upon when you visit Umpqua, you may also find
yourself wandering into the middle of a book reading or a knitting circle.
Special guests of all kinds show up. Once, they brought in a pet psychic.

Most banks have rope lines leading to teller windows; the idea is to move
you along, like an item on an assembly line, until your money has been ex-
tracted and you can be shown the door. But Umpqua is designed to seduce
you into hanging around. The bank has a spacious lounge area with soft
furniture where you can watch a flat-screen TV or use the free WiFi service
while sampling Umpqua's own blend of gourmet coffee. At some point, if
the spirit moves you, you can do some actual banking at a concierge-style
teller desk—and after each transaction, you're rewarded with a chocolate
coin served on a silver platter.

This small bank—which has done such a good job of reimagining the
banking experience that researchers from larger banks now pay money to be
allowed to come into Umpqua and study the way it works—is a product of
design thinking. Its chief executive, Ray Davis, is not a designer but he intui-
tively employed design principles when he set out to re-create Umpqua back

in the late 1990s and early 2000s. Davis, who ran a bank consulting firm before being recruited to fill the top spot at Umpqua, began by asking stupid questions such as *Why do people come to a bank?* and *What does a bank do?*

He teamed up with design researchers and used empathic study sessions to get a better understanding of what people really thought of banks. Some of the findings were not surprising: Yes, people hate the fluorescent lighting and the rope lines, to say nothing of those damn chained pens. But, even in the digital era, people seem to like the idea of their money being in a building they can drive past at any time. And they also like to be able to see the vault (they realize their money may not actually be in there, but it still comforts them to look at a big sturdy lockbox). Davis and the design researchers at the Ziba Design firm of Portland learned something else: A lot of people don't trust big banks.

The research showed that people felt disenfranchised and disconnected from the large financial institutions and were more inclined to trust a small community bank. This was well before the subprime mortgage crisis and the big-bank bailouts, yet the people Ziba studied already had an uneasy feeling that perhaps the large banks might not be doing the right things with their money. "And as we now know," says Ziba creative director Steve McCallion, "they were right to feel that way."

To appeal to these "localists," Ziba suggested that Davis play up the community hub aspects of the bank as it opened its new Portland flagship "store" (Davis doesn't like to call them "branches") in 2003. "We wanted to shift the notion away from banking being all about speed and efficiency and getting you in and out the door," says McCallion. "We decided to focus on the idea of 'slow banking,' with people you know and trust."

AS HE remade the bank, Davis and his designers began to jump fences, taking ideas from other seemingly unrelated fields such as the hotel industry and importing them to his bank. And as they did this, they were often guided by the use of metaphors. McCallion's firm is a big believer in what he calls "transformational metaphors," wherein a company thinks of itself as something other than what it is—which, in turn, changes the way it behaves. If a company thinks of itself only as a bank, it behaves one way. But if it begins to think of itself as a boutique hotel, or as a country store, or as a community center—and Ray Davis had all of these metaphors floating around in his head—then it starts to behave another way.

Thinking metaphorically is common in design. Bruce Mau hit upon the idea that businesses might benefit by thinking that way in the mid-1990s, when he began working with corporate clients that, Mau felt, were too restricted in how they viewed themselves. One of the laws Mau included in his manifesto was, *Work the metaphor.* He declared that every product, every service, every brand "has the ability to stand for something else." And if a company could begin to think this way, it could have a liberating effect by opening up new possibilities, offering fresh ways to present that company and its services to the public.

Davis did not know Mau from Mao, but he "worked the metaphor" like nobody's business. He turned Umpqua into one part Ritz-Carlton, a little bit of Pier 1, a touch of Elks Club, and threw in a double-shot of Starbucks.

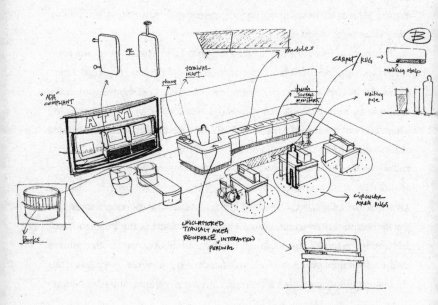

The bank also used metaphors to breathe life into mundane bank products. Bank accounts were named after life-stage aspirations, such as "Go," "Reach," "Savor," and "Cruise." And some of the services were packaged in offbeat ways: If you want to switch your account from another bank to Umpqua, you get your application from what looks like a soft drink can (the cans are in vending machines that instruct you to push a button "for a refreshing change of bank").

One of the more clever metaphors dreamed up by Umpqua involved an outreach program to small businesses. To draw attention to the loan program, Umpqua set up a parallel program geared to local children, inviting them to apply for a lemonade stand start-up kit from the bank. "We felt that the lemonade stand was a perfect metaphor for small business, because in many ways it represents the original small business," says Jim Haven of the Seattle design firm Creature, which helped create the lemonade program for Umpqua. Creature designed thirty top-notch lemonade stands (collapsible, on wheels, made of aluminum, and quite stylish looking). To apply, kids filled out an application form written in crayon, designed by Creature to run in local newspapers. Those applicants who won a stand also received ten dollars in capital and a starter kit with tips on launching a business. The lemonade stand program made a big splash in the community and was all over the local news. Simultaneously, it drew a lot of attention to Umpqua's small business program—and lured in more than two thousand new business customers.

From a bottom-line standpoint, design helped Umpqua grow organically—and without having to take the financial risks that got so many other banks into trouble. When Davis took charge of the sleepy regional bank, it had $150 million in deposits, and he was able to take that up to more than $7 billion, with nearly 150 branches/stores. Like everyone in the banking industry, Umpqua has taken its lumps during the financial downturn, but its overall performance has remained steady, even as other small, independent banks have collapsed. The slow-local-trusted angle has become a strong selling point in turbulent times.

Davis thinks of what he's done as an attempt to "redesign the delivery system of bank services." Umpqua's still doing what banks do, but it's getting those services to the customer in a different way by creating a very different experience inside the bank. In the process of doing this, Umpqua is also designing a complex and evolving relationship with people who now come to the bank for the personal service, the activities, the sense of social connection. While other banks try to compete on, say, CD rates, Umpqua has given people a variety of reasons to keep coming, most of which have nothing to do with rates.

It's worth noting that Davis happens to love what he does, in part because he doesn't see himself as a banker but more as a showman, concierge, community organizer, proud shopkeeper, and orchestral maestro. He doesn't really see himself as a designer, he says. But he should.

THE SUCCESS of a company like Umpqua is normally held up in the business press as a stellar example of "customer service"—just a matter of trying harder to be nice to customers. But the reason it's so hard for those visiting "spies" to duplicate what Umpqua does is because Davis is doing something that is simultaneously highly creative and maddeningly complex. He is not just handing out chocolate coins on a platter. He is mixing performance art with systems design. Davis has figured out that his business is about more than mundane banking transactions—that there's a bigger, more interesting story and experience he's trying to bring to life each day.

A more common term for what Umpqua is doing is "experience design," a concept that has been circulating in the business world since the 1990s. But only an elite group of companies—including Apple, Target, Virgin Airways, and on a smaller scale, others such as Umpqua—seem to have mastered it. What's required is a design-driven business approach that tends to start (as it did with Umpqua's Davis) by reexamining basic assumptions about what a company does and what that means to the outside world. Also critical to the process is the principle of "going deep" to try to get a sense of how people are living their lives and how certain products or services may fit into those lives.

As design researchers have delved deeper into the ways people live and interact with products and services, there's been a growing tendency to think in terms of the "experience" people are having as they engage with commercial offerings. This might seem to be an obvious thing for companies to think about, but it isn't. The more natural tendency is to focus on the products themselves (the buttons, bells, and features) rather than on the actual experiences of the people using them.

In making this change, designers had to shift perspective. Instead of looking at their creations from the creator's perspective, they had to try to imagine the end user's viewpoint. Technically speaking, no one can design an experience for you, since the perception of the experience happens inside each individual's head. What designers are trying to do is orchestrate the variables likely to shape the way the experience of using a product or service plays out for you. To better understand all those variables, designers had to begin to get out of the studio more—which is why the rise of experience design is closely linked to the increased use of the empathic research.

Through direct observation, designers began to see and better understand the full range of activities that might be associated with a given tool or device, including small details that sometimes make a big difference. OXO thought its measuring cups were just fine—all you need is a cup with

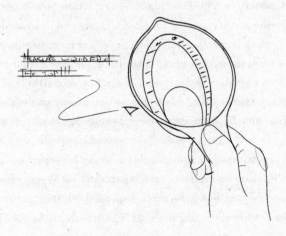

measurement lines, right?—until it became clear, from watching, that when people were filling the measuring cup at the sink, they constantly had to bend and contort their bodies to be able to see the fill level of the cup, which couldn't be seen from above. This led to the angled measuring cup, which enables you to see the fill level when looking down from above.

Michelle Sohn of OXO says the funny thing about the measuring cup story is that people using the old cups never even realized they were doing all that bending—to them it was just part of the normal experience of using a measuring cup. It was up to OXO to study the range of the experience and see that there was something going on, about midway through, that wasn't quite right and was in need of improvement.

Another way of thinking about the design shift from *object* to *experience* has to do with bringing the element of time into the design equation. Designers began to think temporally—to consider the experience someone might be having over time. The old way of thinking was that you designed something in your studio, and then you were done with it. But as Smart Design's Tom Dair explains, "When you start to think in terms of experience design, the designer must step back and envision a long sequence of events someone goes through as they interact with your design." This sequence actually begins well before the item is even purchased, then continues through the critical time when someone brings a product home and opens the package (designers call this the OOBE or "out-of-box experience"), and then keeps going as the product is used over time and becomes part of someone's life. In this cycle, the job of design really never ends—not even after the product dies (that's when issues of designed sustainability and recycling come into play). This represents "a higher level of design," says the IDEO cofounder David Kelley. "There may be things that are not directly part of the product itself, but you still need to design them, because they're part of the experience."

Technology has made it more possible to wrap a full experience around a product. The Nike Plus shoe is a good example of what happens when you ratchet up the "experience quotient" on something that starts out as a fairly simple object: a pair of running shoes. As Nike delved deep into the lives of runners and started to think more about the larger experience, the design challenge became more and more complex. Runners didn't just need arch support, they needed to be stimulated, to monitor progress, to socialize with other runners. The only way to deliver such a holistic experience was to think outside the shoe box and design what is really a system, not an object.

Nike approached the challenge by first seeking out partners, collaborating with Apple and also with high-tech designers such as the firm R/GA, known for interactive Web marketing. What resulted was a system that starts with a chip embedded in a shoe, which wirelessly links to an iPod, which in turn connects to a Web site—where the shoe's owner can track personal running data and progress while also interacting with other runners who are part of a Nike Plus community. Being part of that linked group allows runners to compare performances, find running partners, and share motivational tips.

Nike worked with R/GA to build and host what is essentially a mini Facebook for runners, and while it's a more complex job than creating another ad, the payoff is potentially huge (Nike has already seen double-digit growth in market share since the system was introduced).

This designed social network helps Nike to expand and enrich the experience associated with its product, while also enabling the company to maintain constant contact with customers as it becomes part of their daily routines. It's a level of engagement and influence that advertising can't even begin to approach, says R/GA chief Bob Greenberg, who has been working on similar social network approaches for Nokia and other clients. In one way or another, "every company needs to think about doing what was done with Nike Plus," Greenberg maintains—which is to say they need to design ways to make their products and services more interesting, relevant, and useful to people in their daily lives.

As you move from a boxed product like the Nike Plus to a larger service such as Umpqua banking, the challenges of experience design become even more complex. Umpqua must take into account all the possible points of contact (aka "touch points") between the bank and the public and then try to orchestrate all of those encounters to the greatest effect. If that orchestration hits a false note at any point—if Umpqua's homemade gourmet coffee is bitter one morning—the whole experience begins to unravel, as does the "story" that Umpqua is trying to write in the minds of its customers. But how do designers figure out what makes for a compelling experience in the first place?

5.2 THE "COMPELLING EXPERIENCE FRAMEWORK" (OR WHY A PIRATE ATTACK IS MORE ENGAGING THAN FIFTH-GRADE MATH CLASS)

Designers love to chart and map things, so it's not surprising they've tried to make maps of human experience. "Experience mapping" generally consists of plotting the progression or stages people go through as they engage in various activities (which is not to be confused with "emotional mapping," a design tool used to diagram human feeling—again, designers will map just about anything).

An elaborate experience map was produced, at considerable expense, by the Chicago-based design and innovation consultancy group Doblin Inc., which commissioned an exhaustive research study that tried to deconstruct a single human experience into a clearly defined set of stages. The result of this research was the Compelling Experience Framework, and it posits that there are five phases of a consumer experience: attraction, entry, engagement, exit, extension.

Doblin's framework was based in part on sociological theories about what makes ritual important to human beings. Whether in a ritual or a consumer experience, there must be a progression that starts with attracting and drawing people in, then proceeds to deep engagement (characterized by a sense of losing track of time), and that finally concludes in a manner that is distinctive and has a lasting impact. Overlaying these time-based stages are six intensity-based attributes, which can be adjusted to make an experience more compelling. Doblin settled on these after studying thousands of expe-

ATTRACTION	entry	ENGAGEMENT	exit	EXTENSION

| defined |
| fresh |
| immersive |
| accessible |
| significant |
| transformative |

riences that people characterized as being compelling. What emerged as the most commonly agreed-upon compelling qualities were "defined," "fresh," "immersive" (meaning you lose yourself in the experience and time fore-shortens), "accessible," "significant," and "transformative" (meaning the experience results in a change that is clear and recognizable).

According to Doblin's president, Larry Keeley, designers must consider all of these stages and attributes as they try to orchestrate an experience. Traditionally, product design has tended to focus more on the up-front stages—the emphasis was on attracting you, getting you to buy—and less on the back end. That was short-term thinking, Keeley says; the back end is where you create the lasting effects of an experience, and thus it is the point at which customer loyalty is engendered.

The Doblin group created this framework model for its business clients, but Keeley maintains that it applies to everything from visiting a hospital to attending an educational class: Each is an experience that has the potential to engage people and perhaps even transform them in some way. However, if the experience is not compelling—Keeley points to the shortcomings of a typical fifth grade math class when it comes to engaging students, and particularly girls—it may squander that potential.

There are many examples of well-designed compelling experiences. Keeley likes to cite Chicago Cubs baseball games in historic Wrigley Field, whose unique ivy-covered walls and intimate environs make the games "compelling even when the Cubs lose," Keeley notes. But for sheer excitement and lasting impact, the Cubs can't beat the pirates—not the ball club, but the real thing, circa 1700s. The designer Brian Collins points out that everything one needs to know about designing a compelling experience can be learned from Blackbeard.

"Just imagine yourself back in those times, sailing in a Spanish galleon on the Caribbean," Collins says, adopting the tone of storyteller and sketching an image as he talks. "You look out and notice another ship in the distance. When you peer through the telescope to get a better look at that ship, you see a flag flying. As it gets closer, you can make out what's on the flag—a skull and crossbones. As soon as you see that symbol, you know exactly what kind of an experience is in store for you."

As Collins explains, that pirate "brand" had a story behind it that everyone knew, and the story was built and reinforced by memorable experiences—all of the previous legendary encounters that had taken place between the pirates and other ships. The pirate experience had each of Keeley's stages, with all the intensity attributes dialed up to full blast. And everything about it was symbolized in that designed logo, which sent a clear message to all who saw it. "It was a brand promise," Collins says, "and the promise was: *You're f****d.*"

Each time the pirates delivered on that Jolly Roger promise by creating a compelling encounter—complete with pistols, swords, axes, and various other terrifying accessories, leaving you with scars and tales to tell—they added to the legend and they solidified what that brand meant in people's minds. Like any good experience designer, the pirates understood the importance of theatricality. They decorated their ships and sometimes themselves for added effect (Blackbeard used to braid burning hemp fuses into his hair so that he'd look like a man on fire).

The pirates were so adept at designing experiences and living up to their brand's promise that eventually, Collins notes, all they had to do was wave that logo flag and other ships would drop their cargo into the water and flee. Thus, the pirates managed to do what every brand marketer dreams of—get people to hand over their loot without resistance.

BRUCE MAU first began to think about experience design when he was creating Zone books. In some cases, he saw the book as a starting point that could be expanded into a much broader experience; what began as a book might morph into a series of posters, and then end up being a mini film festival. But he also treated the book itself as something that could be designed to create a richer, more immersive experience. "I tried to design it so the experience correlated to the subject," Mau explains. A book about cities was designed not just to illustrate urban qualities but to model those qualities, "to behave as a city does—with friction, jump cuts, congestion, things moving in opposite directions."

Mau even tried to design the experience of how you look at, read, and engage with a book. He wanted his books to be complex but not cluttered looking, so he designed in levels. On the surface level, his designs seemed fairly clean and simple. But as you spent time with the book and looked at it more closely, patterns of complexity emerged. In one book, the structural patterns of a musical composition were embedded in the graphic design. This was by no means obvious, but even if readers failed to see exactly what was going on, Mau felt that they "would realize, as they went through the book, that there was a coherence at work," and so it would add depth to the experience.

As Mau's studio began to move beyond designing his own books and started working on design projects for various large corporations back in the 1990s, Mau brought his expansive way of looking at things into the land of cubicles and, not surprisingly, he felt the squeeze. He was struck by how narrowly these corporate clients seemed to view the world. "What seems to happen in business," he says, "is that in the interest of security, profit, predictability, and sometimes just priggishness, we collapse down the human experience to something simple, that we can easily control." The result was a disconnect: People were living rich, messy, and interesting lives, while companies were focused on sterile policies and banal corporate speak. This made it difficult for companies to communicate to, or engage with, the "real world" in a compelling way, Mau believes. When he tried to convey this idea to one of his clients, Nokia, he sketched the following drawing:

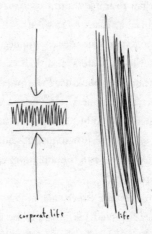

corporate life   life

"I was trying to make the point that if you, as a company, stay within that limited emotional range," Mau says, "then you are cutting yourself off from all the possibilities in the larger world where people actually live."

Mau felt that companies had to do a better job of somehow connecting their business—whatever they happened to be offering or making—with the larger experience of people's lives. One way was to try to instill more meaning into a company's offerings through the use of metaphor. He tried this out with one of his early business clients, Indigo, the leading bookstore chain in

Canada. The company wanted Mau to design a more interesting-looking bookstore. Mau's thought was this: When you enter Indigo, you're stepping into an experience where literature and culture come together on the printed page—you're walking into one great big book, essentially. So he designed the space to live up to that store-as-book metaphor. "We approached the design of store signage as we would a book's title page," according to Mau. "We imagined that a patron, on entering an Indigo store, could effectively 'flip' through its table of contents by scanning the signage."

Mau did something else at Indigo, and that was to try to use design to suggest that Indigo wasn't really selling books—it was selling culture. So throughout the store, literary and cultural references filled the walls (including the "World Needs More Canada" wall, featuring the names of Canadian writers and artists, playing off Canadians' sense of cultural pride). Mau convinced the chain's management that it had to sweat the details of the in-store experience. The store signage, shopping bags, and everything else from the way employees dressed to the way they answered questions about books would either reinforce or undercut this feeling that Indigo was an oasis of literate culture.

Of course, books are cultural objects to begin with, so one could argue that this made Mau's task easier. But books can also come to be seen as commodities, if they're presented as such by an unimaginative bookseller. And the same is true with any product or service offering. Whatever that offering may be, Mau contends that it should be "surrounded by intelligence and wrapped in culture."

Culture and intelligence are transmitted more by signal than by words. People tend to look for clues: *Look at the consistency in everything associated with this brand; these guys seem to have thought of everything; they really get it.* Amid all the noise that people are bombarded with in the consumer marketplace, what tends to stand out is "signal," which Mau defines as "noise with a pattern." A brand identity signal can have nuance and complexity; it can be articulated through color, texture, images, light, smell. "Every gesture and decision a brand makes helps to define that voice," Mau says.

5.3 TRANSFORMING THE "GRUBBY SELF-INTERESTED TRANSACTION"

What Mau was doing in Canada was also being done in the United States and Europe by companies trying to use design to separate themselves from the rest of the field—to say, in effect, *We're not like all the other companies. There's an intelligence at work here; there's a culture that you should be part of.*

Target understood this when it created stores that were idea driven, not price driven. The chain tapped into a growing public fascination with design as it linked up with designers (such as Michael Graves) who were emerging as cultural stars. Meanwhile, in the supermarket sector, Whole Foods used the "transformational metaphor" of an open market or bazaar—which led the company to break rules of supermarket design, with smaller aisles that had areas around aisles overflowing with food samples. "Butt brush" be damned: The Whole Foods design encouraged people to reach across one another and turned the grocery shopping experience into something that felt different—more social and inventive, less of a chore. Starbucks did something similar with coffee shops: The transformational metaphor there was an Italian espresso bar, where people could socialize and while away hours over a cappuccino.

There are good reasons why people responded so well to these designed experiences. Part of it goes back to Donald Norman's "reflective level" of design, wherein people look for meaning and self-worth in the products they own and the services they use. Writing of this phenomenon in the *Harvard Business Review*, Tim Brown of IDEO observed that, in an affluent society, "as more of our basic needs are met, we increasingly expect sophisticated experiences that are emotionally satisfying and meaningful." That craving for rich experiences cannot be satisfied by simple products, Brown notes. What's required is "a complex combination of products, services, spaces, and information."

Target, Whole Foods, and Umpqua used this holistic design approach to elevate what was once just shopping or schlepping to the bank. Commenting on the trend, the *New York Times* columnist David Brooks wrote: "People want an uplifting experience so they can persuade themselves they're not doing a grubby self-interested transaction." In the process, Brooks says, those people will tend to look for "cultural signifiers" that are designed to make them feel they've made the right choice—that they're in good company, and they're part of something designed for people in the know.

The trend toward offering designed experiences may have launched during boom times, but it is no less valid in lean periods. For the companies that are serious about it, it represents more of an evolution than a trend or fad; it has helped raise those companies out of the cutthroat commodity business. And even as the business market has contracted—with Apple, Target, and other design-driven companies feeling the pinch, like everyone else—in some ways, these businesses are on firmer ground than cut-rate competitors (the ground is shifting for everyone, to be sure). The Apples and Umpquas of the business world have built up a storehouse of positive experiences with customers, which translates to brand equity and value. They can attract people with something other than heavy discounts. And their immersive experiences can create a kind of cocoon effect, giving people the sense that even if the world outside is going to hell, the world inside the Apple Store or Umpqua bank is still functioning according to plan.

THAT FEELING is evoked by way of design decisions and details that shape the experience of walking through an Umpqua bank. The wayfinding design that leads you from one area to the next relies not so much on signage as on sensory cues. "You're led through the experience by what you see, what you hear, what you smell—this is what causes you to proceed or to stop and relax," says Lani Hayward, Umpqua's creative director. For example, the music gets louder in the bank's more recreational areas and gradually softens as you move toward the "functional" areas where the actual banking happens.

Ziba Design's Steve McCallion says that various design elements inside the bank are intended to stir "reminiscent" feelings without tipping over into nostalgia. Which is to say, you should feel that some of the things you see at Umpqua are vaguely familiar, without knowing where you've seen them before.

For instance, while the bank design is guided by those "boutique hotel" and "community hub" metaphors, none of that is stated outright—the idea is to let people unpack it for themselves. The design does provide clues, however. One of the most powerful can be found in the teller area. Located around a corner from the entrance so it's not immediately in sight (Umpqua believes the "entry" stage of the compelling experience should start with pleasure, not business), the bank's teller row, as previously noted, is designed to look like a hotel concierge desk—it's a smooth, unbroken bar, without the multilevel cuts and dividers that are usually built in to a teller row.

To make this work, Ziba and Umpqua had to address some practical design issues regarding the teller desk. The cuts and dividers in a typical teller desk are there for a reason: They allow the teller to count the money on one side, in view of the customer, and then slide it across the desk, with the dividers on either side providing privacy from the person next to you.

On the new desk, Ziba addressed the privacy issue by using a series of elongated desktop lamps, which served to block the view of your neighbor while enhancing the "hotel desk" feeling. But there was another matter that had to be resolved. The concierge-style desk didn't provide a good place for the teller to count the money (the teller was working at a different level and would have to count the money out of sight, then lift it to put it on the counter; tellers worried that they'd be accused of shorting people). So Ziba reintroduced the concept of the money tray—an old banking custom that had been eliminated over the years for efficiency purposes. The teller counted the money on a separate tray, which could then be handed to the customer. And to add one more small hospitality reference, it was decided that a chocolate coin would be put on the tray along with the money. So the tray actually serves a highly functional and practical design purpose, but when combined with the chocolate, it seems more like a personal touch from an earlier, more thoughtful era.

McCallion observes that people get a good feeling from the smooth teller desk and the money tray, even if they can't quite put a finger on why. "With transformational metaphors like the hotel references," he says, "you can tap into the meaning that somebody might have with something else and bring it into this new context. So it almost becomes a shorthand way of telling the story you want to tell. In a way, we're tapping into what Jung and Joseph Campbell talked about—the idea of the creation of meaning through myths, metaphors, themes."

If people like the old-fashioned tray, they also like the newfangled twenty-five-foot touch screen video wall. This idea was developed in Umpqua's Innovation Lab, which is constantly working up new prototypes to add to the overall experience. Following the classic design methodology, Umpqua envisions new possibilities, builds them, and puts them out on the floor to see if they work. It's critical that Umpqua keep evolving and changing the experience, because in order to design an immersive environment—a place where people become so absorbed that they lose track of the time—the designer has to strike just the right balance between familiar and fresh, between comfort and stimulation.

The psychologist and author Mihály Csíkszentmihályi, an expert on "immersion," says that it can be achieved by an environment (or even by a well-designed product) that allows the individual to feel in control, while also providing challenges and fresh stimuli. In a store setting, immersion can produce a "stop time" effect. Starbucks was among the first to build its business model around this type of designed experience. One of the keys to Starbucks's unlikely business plan—which aimed to sell one of the most readily available commodities, coffee, for three or four times as much as everybody else—was to offer a designed oasis to go with the coffee. Starbucks wrapped culture around the coffee cup, with its savvy baristas and stylish furnishings; the air was filled with irresistible smells and a contemporary sound track. Most important, Starbucks encouraged hanging around, while most every other business was still following the "move 'em in, move 'em out" model. Starbucks made a conscious and designed effort to stop time.

So what went wrong? The company's well-documented troubles over the past couple of years are usually blamed on overexpansion, but to be more precise about it, one could say that Starbucks lost the ability (or the willingness) to stop time. As the chain became more popular, particularly among those in need of a quick caffeine fix, the hustle and bustle took over. The shops became associated with people rushing in and out or waiting impatiently in long lines. That languorous Italian espresso bar experience was disrupted; Starbucks lost the metaphor.

TO AVOID that fate, Umpqua must protect the integrity of the experience, even as it adds new wrinkles to try to keep things fresh. The designers at the firm Adaptive Path, a leading expert on designed experiences, say that it's critical to make sure a company experience aligns with one or two key driving principles. The principles should serve as a "North Star" to guide experience designers as they try new innovations.

At Umpqua, Davis has a couple of North Stars that guide him. One is customer service—which has been something close to an obsession of his from day one. Boutique hotel–type service is at the core of the Umpqua experience, and it's central to the design of the operation. For example, when you walk into any one of Umpqua's banks, you see a special phone that's kind of like the hotline to Batman, except this one rings direct to Davis's office; any customer can call the president of the bank, anytime. "They push a button, I pick up the phone—that's no accident, that's design," Davis says. (The bank knows that, fortunately, most people won't actually use the special phone, and that's factored into the design, too).

Davis's attempts to follow his "service" North Star can lead him to try just about anything. Once, upon visiting a Ritz-Carlton hotel, he was surprised when people at the hotel front desk greeted him by name before he even had a chance to say who he was. How'd they know? Davis discovered that the hotel used walkie-talkies—the bellmen peeked at luggage tags while greeting you outside, and then they'd radio your name back to the front desk. Davis immediately went out and got waistband radios for the people working at Umpqua bank.

In addition to service, "community" is the other North Star for Umpqua. Most of the new features added to the experience—one of the latest being a Nintendo Wii bowling competition, right inside the bank—are designed to foster the sense that Umpqua is a place for local people to get together and share the experience. It's interesting to contrast the way Starbucks added music sales to its offerings with the way Umpqua did so. In Starbucks's case, the music being offered for sale was somewhat generic pop music, repackaged by Starbucks; in Umpqua's case, it was original music from local bands, which provided a great fit with the "community" experience.

Designers must sweat details to ensure that nothing disrupts the gestalt of a designed experience. Often, less is more (a lesson Starbucks should have heeded, as it was adding more and more offerings to the mix). Nothing should get in the way of those North Star objectives.

GOOGLE, FOR instance, goes to great lengths to design what appears to be a very simple search experience. Google's designed experience may seem far less immersive than Umpqua's, since it's happening entirely on a computer screen, but Irene Au, the company's director of experience design, says that the company has a design challenge not unlike that of a Starbucks or an Umpqua. Its design must attract and engage and, above all, not get in its own way. Google's North Star is very clear: Its goal is to deliver information in the fastest, simplest way, and everything about the experience design flows from that.

When Google started, the company's engineers tended to design from the inside out, meaning they relied heavily on their own instincts and judgments to design what they felt would be the best user experience for everybody else. Those instincts proved to be dead-on, but nevertheless the company gradually has adopted more of an outside-in design approach. Google now relies much more on watching and studying people as it tries to figure out how to design a better experience.

Often, what Google is looking for are obstacles that can be removed. "With the user interface design, it's often a matter of simplifying the elements on the page so that we don't have extra adornment or too many functions." It's not an issue of aesthetics, Au says, but "more because we don't want the interface to get in the way of our users getting exactly what they want the moment they need it."

Google's designers are ruthless when it comes to streamlining; even the most minor elements on the page are tested, and survival is dependent on being deemed useful by a given percentage of users. The density of information on the page is designed in accordance with people's behavioral patterns on the Web. "For a long time people didn't want to scroll and even now they prefer to see what's most relevant at first glance," Au says. "Part of it is basic human factors, and it's very logical." But the Google experience also can be surprisingly emotional, in terms of, say, how people feel about certain colors. If Google changes the color or even the shade on its header bar, "it can have an impact on how many searches per day we have." Overall, in terms of its design the search experience is quite fragile; "even pixel-level changes have a dramatic impact on usage," Au observes. "It's the kind of thing where you move things two pixels this way or five pixels that way, it can mean plus or minus a million dollars in revenue."

We tend to think of companies like Apple and Google succeeding based on technology, but their real success often lies in experience design. That's where some of the most critical innovation takes place, notes Peter Merholz, president of Adaptive Path. Merholz points out that, typically, technology makes something new possible—which is fine until competitors quickly copy the technology, and then "featuritis" sets in as the various players try to one-up each other with minor feature upgrades. But at some point, according to Merholz, what's required is "a quantum evolution—beyond technology and features—to the satisfaction of a customer experience." Google didn't invent search engine technology; it simplified and improved the search experience. And Steve Jobs did not invent the MP3 player; he designed a better, richer experience around the technology.

By doing that, says Ziba's McCallion, Jobs also managed to design a halo for himself and his company. "You know, Microsoft wonders why people get so upset when the Windows system crashes, and they try to point out that it crashes less than Apple," McCallion says. "But try telling that to an Apple person, and they'll vehemently deny it. And the reason why is that Apple has been able to create a halo of good experience around its products. That's what we're doing with Umpqua—trying to create that halo. If Umpqua has that halo, then maybe in times of difficulty, people will hang with them longer."

5.4 HAVE THE "MAD MEN" LOST THEIR MOJO?

That halo could be thought of as a form of advertising, except that it's vastly superior to advertising. What happens inside Umpqua's bank is more eloquent and compelling than any sales pitch the company could write. As for Google, well—when was the last time you even saw a Google ad? Or one for Facebook? Apple does advertise, but, as Brian Collins notes, what's interesting is the content of those ads. Many of them simply show a close-up of the product, with a hand pushing buttons as the experience unfolds on an iPod or iPhone screen. No scripts, no hype, no sales pitch necessary: Apple lets the product and the experience do the talking.

Collins says all of this can be taken as evidence that the experience design movement is reinventing the marketing model that has been dominated by ads for the last half century. That old model made it feasible that a large company could offer a lackluster customer experience, yet still coax people to purchase its bland offerings, thanks to the sheer power of message bombardment. But over the past decade, a flip-flop began to take place. Ads started to lose their power because of changes in media (including more fragmentation and greater audience control and participation). Meanwhile, those same changes in the media heightened the importance of providing quality customer experiences because customers now had more ways to talk to each other. Today, increasingly, the experience itself *is* the advertising.

Dan Formosa, a design researcher and partner at Smart Design, explains it this way: "If I'm a customer in the store nowadays and I see something that interests me, my first reaction is to think, 'I've got to go online and find out what people are saying on Amazon about this product.' And what you get on the Web are people's experiences with a product or service—you don't get a sales pitch or slogan, you get the truth of how that product or service is performing in the real world." The bottom line, Formosa says, is that in today's marketplace, the best customer experience wins.

The companies that are unable to figure out how to design and deliver that experience have little else to do but make pleas for our attention that are mostly ignored—in other words, they advertise. Collins believes we may now be reaching the point at which advertising becomes the penalty paid by companies that cannot design well. "In this new environment," he says, "you could think of traditional advertising as a tax on laggards."

IF DESIGN really is the new advertising, this can be seen as good news for innovative companies. It means they can design their way into the public consciousness, even though they may lack big ad budgets. A pioneering example of this was the launch of the Mini Cooper automobile in the U.S. market back in 2002. At the time, the Mini faced a daunting challenge. Never mind the fact that, at the time, America was mad for sport-utility vehicles, the bulkier the better. The bigger problem for Mini was that it had a small marketing budget by car industry standards. Successful launches of new car brands in the United States typically called for a couple of hundred million dollars in advertising. Mini had less than a quarter of that amount to spend. But it did have a very creative marketing manager at the time, Kerri Martin.

Martin believed that a big, splashy TV ad campaign wasn't really necessary, or appropriate, for the modest Mini. She felt the car could succeed in large part on the strength of its own design. It handled well, was surprisingly roomy inside, and was about the cutest thing on four wheels. Mini's vehicle design team in Germany (the brand may seem quintessentially British, but it's owned by BMW these days) had taken great pains, according to top designer Gert Volker Hildebrand, to give the car's body a slight wedge shape that suggested forward movement. Hildebrand's team spent even more time manipulating the interplay of hood, headlights, and grille to give the car the friendliest "facial expression" possible (even with all that effort, Mini spends less than 1 percent on design—and yet customer surveys show that design is why 80 percent of people buy a Mini).

Since the car offered such a well-designed experience, Martin had the radical idea that perhaps the launch could, itself, be designed as a unique experience—one that didn't involve big commercials and slogans, but rather a hundred small connections and "touch points" with potential customers.

Early in the U.S. launch effort, Mini brought Yves Behar into the process. This was an unusual thing to do, because Behar's then-tiny San Francisco start-up was known for designing shoes and lamps (this was before the XO laptop), and had no connection to automotive design. But Martin didn't need help with the car design; what Mini needed was an ecosystem, a supporting network of products and services that could help elevate the primary product into a fuller experience. Behar set out to enlarge and expand the Mini driving experience by creating specially designed accessories—driving gloves, a special driving jacket, a wristwatch.

Normally, Behar says, car accessories are generic items with a logo slapped on them. But with each of his accessories, Behar aimed for complete reinvention of whatever the item was, to bolster the larger idea that driving a Mini was a different and unique experience. This was a perfect assignment for Behar, because he tends to want to reinvent everything, even the most basic things; he's the kind of person who often looks at how things are being done all around him and shakes his head. "We have such a learned behavior that gets formed around the imperfections built in to the products we use," Behar says. "I get amused by these idiosyncrasies. Like how we move our elbow to read a watch. This is learned behavior! Human beings weren't born thinking that they would learn the time by bending their elbow. You think, how about people doing it in a more natural way, an easier way."

So Behar designed a Mini watch with a time display that changed directions based on the position of your arm (this was five years before the iPhone had a display that changed direction, Behar takes pains to note). At last—you could tell time without bending your elbow (which may have mattered more to Behar than anyone else on earth, but it was, nevertheless, a pretty cool feature). The watch also had a bracelet that adapted to your wrist and could be instantly pulled on and off without using buckles or clasps. It was sleek, efficient, and maneuverable, like the Mini. The jacket Behar designed for Mini had a map pocket that folded out into your lap, so you could check the map while driving.

While Behar was designing the ecosystem, Alex Bogusky was asked to design the echo chamber. Bogusky, a young, long-haired motorbike racer who worked with a small but growing Miami ad agency, was technically an adman, but he'd come from a design background and he actually kind of *hated* ads. Kerri Martin knew of him because she'd admired a marketing campaign Bogusky had created for another client a couple of years prior, in which he'd applied a radical systems-design approach to an intriguing challenge: trying to get teenagers in Florida to stop smoking.

5.5 WELL, IF YOU DON'T CARE ABOUT DYING, WHAT *DO* YOU CARE ABOUT?

Bogusky's "Truth" antismoking campaign is famous mostly within the ad business, though it's actually a stellar example of the power of design to do what was previously thought impossible—which, in this case, involved getting teenagers to behave rationally. Bogusky had been approached by the state of Florida because teen smoking rates were on the rise in the late nineties and the state's health department wasn't sure what to do. Traditional advertising appeals—ads saying, basically, *It'll kill you, kids*—were not working. Bogusky immediately started asking stupid questions, such as: *Why don't teenagers care about the possibility of dying? And if they don't care about that, what do they care about?*

Bogusky's team did a deep dive on the streets of Miami, hanging out with teens, documenting everything with video cameras. The agency also pored through teen psychology literature and studied ad campaigns, both for and against cigarettes. There were some interesting epiphanies, including this quite disturbing one: "One of the things we realized," Bogusky says, "was that if you actually set out to design a dream product to market to teenagers, you couldn't come up with anything more effective than a Marlboro cigarette."

That's because the product tapped into 1) a teenager's need to establish an identity; 2) the desire to be associated with distinctive brands; 3) the urge to rebel; and even 4) the normal adolescent eagerness to take physical risks and confront danger. In this context, the danger associated with smoking actually made it "sexier."

So if they didn't care about dying, what *did* they care about? Bogusky's researchers watched and listened and came up with this answer: They cared about being manipulated, used, "played." This led to Bogusky's glimmer moment: Why not take all that pent-up teen angst and direct it against the shady sales tactics employed by the tobacco industry? Why not turn that industry's own marketing muscle against itself?

Bogusky created a brand name, Truth, and designed a logo with Helvetica type inside an oval. Then his agency launched what was not so much an ad campaign as a popular movement—fueled by various designed tchotchkes including leaflets, fliers, stickers, hats, and buttons, distributed at concerts and other teen events. The campaign also disseminated information about the misdeeds of cigarette makers, citing examples of cover-ups, phony ad claims, and outright lies. Bogusky staged pranks, such as having teenagers show up at tobacco company headquarters with bullhorns and body bags. The campaign was extremely cost efficient because, in many cases, the kids who got caught up in it carried the message to others, for free. The whole thing spread like wildfire throughout Florida, and by 2002, five years into the campaign, smoking among middle and high school students in Florida had declined an average of 38 percent.

WHEN MARTIN came to Bogusky, she was hoping he could create a similar type of grassroots movement on behalf of Mini. What she needed for Mini—and what Bogusky had done with the Truth effort—was to orchestrate a designed marketing experience, one that would surround people instead of shouting at them; one that would wrap a certain culture and intelligence around the product. The eventual Mini campaign did all of that in spades and became a textbook example of what advertisers lovingly refer to as "integrated marketing."

To promote the Mini, Bogusky's agency, Crispin Porter + Bogusky, created games, paper cutouts, billboards that seemed to come alive, and cartoon books. It turned heads by using the Mini car as a prop in various live stunts, such as by attaching a Mini to the roof rack of an SUV (with a sign saying, WHAT ARE YOU DOING FOR FUN THIS WEEKEND?) and driving it around towns across America. At sports events, Mini car seats were installed in place of ripped-out stadium seats. Outside department stores, the agency replaced some of those quarter-a-ride mechanical ponies with miniature Minis. At airport terminals, oversize props designed to look like giant pay phones or garbage cans were installed alongside a poster of the Mini with the headline, "Makes everything else seem a little too big." On the Web, the agency created humorous phony Web sites that quickly went viral—including one dedicated to disseminating rumors and blurry photographs that purportedly documented the existence of robots made from old Mini car parts.

Bogusky also joined in Behar's effort to add little extras to the Mini owner's experience. He created an "unauthorized owner's manual," sent by direct mail to each owner after the purchase, filled with both practical and humorous tips. CP+B even got involved with rewriting the lease agreement on Mini's convertibles, inserting tongue-in-cheek language that suggested users were contractually obligated to keep the top down whenever possible.

The Mini campaign had a kind of carnival effect in that it seemed to be going on all around as a living event; you never knew what was coming around the next corner. And all those designed artifacts and messages seemed to connect with one another to form a cohesive statement about the brand. The campaign more than quadrupled public awareness of Mini virtually overnight and created a customer order backlog in Mini's first year on the market. Nobody seemed to notice the lack of commercials.

Except other marketers, who noticed and wondered if Bogusky might be able to save them a lot of money on commercials, too. Burger King, Microsoft, Coca-Cola, and other big brands all came calling on CP+B after that. The Mini campaign didn't kill the thirty-second commercial, but it showed a lot of marketers what life after the thirty-second commercial might look like. It looked complex, diverse, multidimensional, experimental, experiential, and systematized. It looked like design.

HERE IS the bottom line, according to Brian Collins: If advertising is a promise, design is performance. For much of the last century, the business world relied heavily on promises. But you can only promise so much, for so long, before the time comes to deliver.

Mau thinks it's a wonder that advertising has lasted as long as it has. "If you think about the concept of advertising," he says, "it comes down to: 'Let me interrupt you, so I can tell you something that is probably irrelevant to you.'" That never made much sense, Mau says, but while advertisers controlled the media, they could do as they pleased. That's all ending now, and the challenge for businesses is to find a new way to connect with the public. "We have to eliminate interruption and increase relevance," he says.

One of the ways that companies can do that, according to Mau, is to change the way they behave. Mau's thinking goes like this: If a company begins to do more things that are actually relevant and interesting to people, then that begins to take the place of advertising. And if you're doing worthwhile things, then it's okay to tell people about it. "Because if you're telling me something that's relevant and interesting to me, then it's no longer an interruption."

And how does a company ensure that the things it does are perceived as relevant, interesting, worthwhile? Simple, says Mau: You must design everything you do.

6. DESIGN WHAT YOU DO

Can the way a company behaves be designed?

6.1 "YOU CAN'T SELL ANYTHING TO DEAD PEOPLE"

One morning in June of 2008, Mau sat on a stage in the ballroom of New York's Grand Hyatt hotel, accompanied by a senior vice president of Coca-Cola, Marc Mathieu, and a moderator. The event was a design industry trade show, and this particular session focused on Coca-Cola's attempt to redesign its business to be more sustainable.

Usually when corporate executives talk about sustainability, the challenge for listeners is to sustain interest. But Mau is good at presentations, as are many designers, and he livened this one up with dramatic gestures (at one point he held up his drinking glass and asked the audience, "What happens after this breaks?"), along with compelling images flashed on an overhead screen.

One of those images showed a sea of plastic bottles—it looked as if thousands of them had been dumped into the water. Then the view pulled back and it was more like hundreds of thousands. Then it pulled back again and it was more like millions. The image looked real, but it was a digital creation, which Mau had borrowed from a photo artist. This trade conference wasn't the first time he'd shown it; Mau had employed the same image very effectively in one of his early business meetings with executives at Coca-Cola. He'd wanted to give the company a jolt of future shock, so he presented the digitally doctored images but also backed them up with real-world numbers that he'd projected for the next fifty years.

"We showed those two million bottles," Mau recalls, "and then I said to them, 'What you're seeing here? That's nothing. Over the next fifty years, you will leave 2.2 *trillion* discarded bottles in the environment.'" And all those empty Coke bottles piling up all around, Mau said, would become a kind of anti-advertising for the brand. That image, shown in that meeting, helped secure a consulting assignment for Mau and led to his sitting with Mathieu at this conference.

As the design conference wound down, the panel took questions from the audience, and one questioner, a young man with an earnest manner, seemed poised to spring as he took the mic. He pointed out that Coca-Cola really isn't very good for people; that it doesn't contribute much to the health and well-being of the planet. Therefore, he wanted to know, "Why not just stop making Coke? Why not start purifying water instead?"

The Coke executive, Mathieu, seemed taken aback. How do you respond to someone who suggests that you might want to, you know, just put yourself out of business? But Mau was eager to answer because he felt the questioner was getting at something central to the whole discussion of business, consumption, and sustainability.

"I hear what you're saying," Mau said to the questioner, "but I don't think 'no' is the answer to these challenges. Denial—just telling people you can't do this, you can't do that—will not get us to where we want to be. We have to redesign the things we love so we can keep enjoying them." Then, after a beat: "And I happen to love Coke."

Mau observed later that while the questioner might have seemed a tad green and crunchy for a business conference, he was, in fact, representative of what Coke and every other company is up against these days: a public that is more aware of and concerned about what companies are doing—and one that also has more ability to question and challenge business than ever before.

Companies such as Coke are realizing that they must adapt and adjust their behavior in order to survive this new level of scrutiny—and also just to survive, period. As Mau points out, the business world has a clear stake in sustainability issues if only because, "You can't sell anything to dead people."

OF COURSE, Coke is not going to do what that audience member seemed to want—it's not going to kill off the main product and devote itself to making LifeStraws. Somehow the company must balance responsible behavior with profitable endeavors. The sustainability concerns get mixed in with a hundred other objectives, all of them pressing: the need to innovate, to design distinctive consumer experiences, to streamline operations and maximize efficiencies.

As Mau sees it, all of these issues are connected—or rather, they *should be* connected. They are all part of a complex system that is comprised of the behaviors and actions of Coca-Cola, or of any company. If those actions are planned and executed in a cohesive, creative manner—if they're designed—then the overall system presumably functions better and can avoid breakdowns.

This is where Mau's principle *Design what you do* comes into play. It sounds simple, but it's actually a radical use of the term *design.* Businesses are used to designing what they make (the products); they may also be used to designing what they say to the outside world (advertising and communications). But it's more unusual to think of applying design principles and approaches to the full spectrum of a company's behavior—encompassing everything the company does, including what it does behind closed doors.

If design is associated primarily with style and aesthetics, the notion of designing a company's operations might seem pointless. Who cares what it looks like inside the sausage factory? And why should it be anyone else's business? But Mau's position is that there really is no "inside" of the factory anymore. Changing conditions are calling into question the long-held assumption in business that there are two separate realities: the one that is shared with the outside world (in the form of product offerings, advertising, communications), and the one that is considered private (the way a company actually makes things, operates, treats employees, disposes of waste, and generally conducts itself).

If ever business could get away with leading a "double life," it's far less true today. The Internet has brought an unprecedented level of transparency that allows the outside world to see—and comment on—the way a company performs and behaves. "There is no back office anymore," Mau says.

Mau likes to cite the example of Nike. For a period of time, Mau says, the company's upbeat advertising images of athletic empowerment and progressive ideals were undermined by its own overseas labor practices—particularly as the latter came under scrutiny in the media and especially on the Internet. To make matters worse, Mau notes, nothing ever dies in cyberspace. The complaints and gripes about Nike's labor practices continued to remain just a Google search away, even long after the company had taken steps to improve those policies.

The bigger the company, the more vulnerable it is to this problem. Wal-Mart and Procter & Gamble both spend enormous amounts on advertising, but often their actions—the things they do that may adversely impact local workforces or the environment—get more media traction than the ads. In fact, the ads themselves sometimes serve to call attention to inconsistencies. In one instance, the packaged-goods giant Unilever had to answer for why it was celebrating everyday women with its Dove ad campaign, while another of the corporation's brands, Axe grooming products for men, ran male-fantasy ads that could be seen as sexist. In the past, people might not have connected the dots between the Axe and Dove brands, but in today's transparent culture, chewed over by innumerable blogs, such things get noticed.

It's not just that it's easier for the public to monitor companies, via the Web; it's also that people seem to care more about how companies behave. This has ratcheted up the pressure on marketing executives, as the ad executive Lee Clow acknowledges. Clow, who has helped define the brand images of Apple, Nissan, and others, says, "These days, whatever a brand is doing, somebody out there is going to point to it and say either 'Look what a cool company this is' or 'Look at this bullshit company; they tell you one thing in their ads and then they do the opposite.' I think, in particular, young people actually love doing this—it's recreational sport for them to pick apart brands that are hypocrites."

It's more than just recreational sport, however. The relationship between people and brands seems to have changed. There's more of a tendency to identify with and even to worship the "cool" brands while also holding them up to a high standard of behavior and performance. And people now judge a company not just on its products but also on the larger "experience" that is delivered. In that context, they're looking for a sense that there is a consistent philosophy and a cohesive plan at work. The experience can break down when people become aware of missteps by a company, even if the missteps are happening far away or behind the scenes.

6.2 TEACHING CORPORATIONS ABOUT INCORPORATION

Early in his career, Mau began to consider the idea that everything a business does matters; that every action communicates a message to the world and also has consequences on some level. There were a number of small glimmer moments that reinforced this notion. For example, "One day I saw a truck driver with a big rig," he says, "and I had this moment of clarity where I thought, *That driver doesn't know what's in the truck.* And we allow him to remain ignorant and to say, *My problem is getting whatever is in this truck to the right address. Whether it's a dirty bomb or an order of hamburgers is not consequential to me.*"

Mau saw this kind of compartmentalized thinking as standard practice in business and felt that it allowed industry to wreak havoc on the world. "It led companies to say, *We are only going to express our values when we're communicating—but when we're manufacturing and doing all these other things, we don't have to worry about it because those things aren't visible.*"

While many businesses tend to compartmentalize, Mau, as a designer, tended to look for the connections between things. This was a particularly resonant idea with Mau because he'd grappled with the concept of "incorporation," which was the subject of one of his 1990s Zone books. In essence the book was about what Mau calls "the end of the discrete object," and it proposed that the human tendency to think of the world around us in terms of individual and separate objects is mostly an illusion and a way to help us more easily process information and experience. "But when you really start to apply scientific knowledge to all of this, the truth is that every object is not a separate thing but is incorporated into larger flows," Mau says. Or, to break this down to the short version: Everything is connected.

What made the incorporations idea relevant to business was the trend of companies coming under scrutiny from consumers. Suddenly, it seemed, everyone was starting to do what Mau did that day long ago when he looked at that company truck and wondered what was being transported. Moreover, in the new age of transparency, people don't really have to wonder—they can, in effect, *see* into the back of the truck.

The incorporations concept was still much on Mau's mind when he did the Massive Change show. Parts of that exhibit dealt with biological and environmental issues, and particularly with the ways in which complex ecosystems function. Every organism in a system affects and depends upon what is all around it, which means "there really is no 'exterior,'" Mau says. He thought this notion of "no exterior" pertained to the business world, as well—that businesses had to stop dividing reality into "the company" and "the outside world."

Massive Change proposed that entities or organizations could transform themselves by designing new behaviors that adapted to the change happening in the world around them. Cities could change, water delivery systems could change, transportation could change—and companies could, too. But to do so meant reconsidering and reworking a broad spectrum of activities. In business, that might include everything from the raw materials used to supply chain issues to the way employees are managed. The Massive Change viewpoint was that these separate functions should be seen as part of a completely integrated and thoroughly designed system.

When Massive Change debuted in 2004, the first wave of green marketing mania was just beginning to sweep through the business world and a number of major companies were already starting to think about sustainability issues. Coca-Cola was one of them. The company was developing new marketing programs with a greater emphasis on corporate social responsibility and had put Marc Mathieu (no longer with the company) in charge of the effort. When the Massive Change show came to America in 2006, Mathieu flew to Chicago from Europe to see it. He recalls that he was hooked on the show the moment he stepped into the Museum of Contemporary Art and saw the challenge Mau had posted in giant type at the show's entrance: *Now that we can do anything, what will we do?* "That really resonated with me," Mathieu says, in part because it was a question Coke was grappling with as it looked to the future. Mathieu was also intrigued by Mau's idea that companies could use a systems-design approach to better integrate all of their behaviors. He invited Mau to Coke's headquarters in Atlanta.

AROUND THE same time it formed a relationship with Mau, Coca-Cola was also reaching out to a number of other prominent designers, including Yves Behar and David Turner of the design firm Turner Duckworth. The company knew it had to update and improve bottling and packaging, and not just from a sustainability standpoint. Coke's image had been diluted by too many product spin-offs and a lack of consistency in the way the brand presented itself to the world: There were hundreds of different cans and bottles, and it seemed no two Coke vending machines looked alike.

"The Coke brand had become like cultural wallpaper," says David Turner, meaning that it was everywhere but no one seemed to notice it anymore. By 2005, the company determined that it needed to focus on three key areas: brand identity, user experience, and sustainability. And it also dawned on Coke that the common thread connecting these three areas was design.

Coke actually had a great design heritage, as Mau was well aware. Back in the late 1990s, Mau and Frank Gehry were enlisted by a wealthy family that owned a large collection of Coke artifacts; the assignment was to create a museum to house them all. The museum was never completed, but while he worked on it Mau found himself surrounded by artful signage and posters, elegant soda fountains created by the likes of Raymond Loewy, and those classic contour Coke bottles, once hailed as a great American design icon by Andy Warhol.

As Mau immersed himself in Coke's design history, he was surprised to learn that the company wasn't just creating those elegant artifacts; it was also designing an industry from the ground up. The company drew up detailed plans on how to design corporate facilities right down to the shrubbery. In towns across America, Coke had figured out exactly where its advertising signs should be placed in relation to the town's courthouse and its railway station. "They created design templates for everything," Mau says. Coke also engineered the migration of its product from the soda shop into people's homes (to that end, the company designed those stylish bottles, drinking glasses, and bottle openers in such a way that people would *want* them in their homes).

Coke's emphasis on design fell away in the latter part of the twentieth century, when, like a lot of big marketers, the company came to rely primarily on advertising and not design to lure customers; the jingles took precedence over all else. Only in the past few years has Coke begun to rediscover its design roots. To bring some of the old style back into its bottles and packaging, the company turned to the Turner Duckworth firm, which proceeded to repackage the brand in stunning fashion.

The classic Coke bottle shape was now clad in sleek red aluminum and cleansed of promotional clutter, featuring only a Coke logo blown up so large that when it wrapped around the bottles only part of the logo could be seen from any angle. This was design shorthand for expressing confidence. Coke could show you just a sliver of the logo, knowing that you'd still know whose bottle it was. The aluminum bottles created a stir that hadn't been seen since the contour bottle was in its heyday; they scooped up design awards all around, including at the 2008 international ad festival in Cannes. Turner then extended the look to coolers, billboards, and trucks—trying to bring some cohesion back to the Coke ecosystem. Yves Behar, meanwhile, designed a nifty recycling receptacle for the Coke bottles.

THE BOTTLES and the bins were one thing. Mau, on the other hand, set out to redesign the mind-set of Coke. He wrote and designed a manifesto around the idea of "living positively," with a series of guiding principles for the company. Inspired by the Eames motto about providing "the *most* of the *best* to the greatest number of people for the *least*," Mau tried to get Coke to view sustainability as an opportunity, not a chore. Not that it wouldn't require work: He felt the company had to better integrate a variety of disciplines—manufacturing, purchasing, packaging, product development—so that these separate silos could begin to work together to achieve the overarching sustainability goals.

As many a CEO knows, it can be difficult at large corporations to create cohesion across departments and divisions that are walled off from one another. "What you must try to do," Mau says, "is work horizontally in a business culture that is vertical."

The overall operation of a business can be thought of as a wicked design problem in which there are many interrelated challenges and when you improve one, you may worsen another. For example, a company's product development group may pat itself on the back for using recyclable materials, but in order to obtain that material the company may have to do additional shipping, which can negate the intended positive environmental effects. This situation, which is actually quite common in business today, has been referred to as "the sin of the hidden trade-off" and it's a reason why a lot of green marketers aren't really as green as they claim to be.

Designers such as Mau and IDEO's Valerie Casey, head of a green-design business initiative called the Designers Accord, are trying to get businesses to look at the impact of every part of the operation, not just the actual design of products. For example, Casey points out that "there are huge opportunities for a company to redesign the flow of goods. It may be a matter of asking, 'Okay, why do our dockworkers only fill up the containers one-third of the way?' Maybe there's a very understandable reason having to do with their shifts, and that's something that can be redesigned."

Mau contends that bringing design solutions to bear on these mundane operational issues actually can end up saving money while also improving behavior. He points to the example of UPS, which has redesigned its truck delivery systems so that drivers make far fewer left turns. What's wrong with left turns? They cause the drivers to sit idling in traffic, waiting for lanes to clear. By rerouting trucks and cutting down on those turns, UPS cut fuel emissions significantly—but at the same time, it also reduced its shipping expenses.

Coke was, in fact, already doing a number of positive things, including various disconnected efforts to innovate its manufacturing methods and make them more sustainable. But hardly anyone knew about them, because they weren't coordinated as part of a larger, more visible effort. Mau was impressed by several programs geared to water conservation. For example, some of the company's bottling plants had devised new, waterless ways of cleaning bottles. But many of the company's social initiatives were catalogued in thick binders that few bothered to wade through. "You open one of those books and there's so much in there, described in such a dry way, that you just want to close that binder and go have a beer," Mau says.

6.3 "AN ISLAND OF INTELLIGENCE IN A SEA OF STUPIDITY"

Mau believes that Coke and other companies need to build critical mass behind their efforts to behave well—and that doesn't necessarily mean companies have to do *more* good things. Sometimes companies take on too many small, random efforts. For example, Mau believes that Coke could dispense with that thick binder and boil down those six thousand commitments to, say, a hundred high-impact programs or efforts that would actually make more of a difference in the world—while also bringing more favorable attention to the company.

This is an idea Mau has been discussing not just with Coke, but also other clients including Arizona State University and MTV. Instead of doing a bit of charity here and a bit there, he thinks these organizations should focus those scattered goodwill resources into a more clearly defined set of bigger, bolder initiatives.

What Mau is trying to do is rethink the way companies "do good." The conventional way involves setting up what's known as a Corporate Social Responsibility (CSR) department, which is usually cut off from the rest of the business and focused on donating to various charities or setting up generic employee volunteer programs. Mau, in one of his characteristically colorful turns of phrase, says that he views CSR programs as "an island of intelligence in a sea of stupidity." In other words, the company sets up one little isolated sliver of a department and says, *This is where all our good deeds will be done*. And that is supposed to make up for the company behaving any way it pleases in the other 95 percent of its activities. As Mau sees it, CSR enables companies to do more harm than good.

Moreover, Mau doesn't think these goodwill programs create enough goodwill. Since they tend to give money to the usual causes and charities, every CSR program looks pretty much like the next, as do the self-congratulatory ads intended to call attention to these programs. So what the public ends up seeing is a lot more talking than doing.

Mau thinks a better approach is for a company to demonstrate its corporate values through clear and identifiable actions. If Coke were to pioneer bottling methods that set new sustainability standards, that might be one such action; if it found a better way to truck its products, that could be another; if it took the aggressive lead on tackling a particular social issue, that'd be yet another. But these things ought not be done randomly or willy-nilly, Mau insists. Actions should be designed to fit well with the company's mission and personality, and they should all be woven together in a way that has maximum impact on the world and on the image of the company.

Coke is still in the early stages of figuring all of this out—and the financial pressures exacerbated by the recession tend to take precedence over all else—but the company has taken some significant steps: pledging to recycle 100 percent of its aluminum cans in the United States, while currently helping to finance the building of the world's largest bottle-to-bottle recycling plant. Still, it has yet to embark on the kind of bold initiatives that Mau envisions and is still proposing in his ongoing consultations with the company. "The potential for Coca-Cola to create a powerful social movement and to change the world is enormous," Mau says. But getting that to happen, he adds, is "like turning around a very big ship."

THERE ARE a number of smaller boats that have made that U-turn. The Pedigree dog food brand serves up a good example of how transformation design can work in terms of radically changing the culture, attitudes, and behavior of a company in a purposeful way. This is not one of Mau's clients, but Pedigree did have a pesky, bearded outsider come in and, in Mau-like fashion, start telling the company how to redesign almost everything it does.

The designer in this case was Lee Clow, who is technically an ad executive (albeit one with a strong design sensibility) with the agency TBWA\ Chiat\Day in Los Angeles. Clow has worked closely with Steve Jobs through the years not only on Apple ads but also on some aspects of the design of its corporate culture. When Jobs returned to a struggling Apple in the late 1990s, Clow helped to come up with a company rallying cry, "Think Different," that not only appeared in ads but also on banners and T-shirts inside Apple headquarters. It was the company's own version of a pirate flag, serving to remind employees of Apple's mission to zig while others zag. What soon followed was a remarkable string of design innovations: the iMac, the Apple Store, the iPod, the iPhone.

With regard to Pedigree, Clow had a particular interest in the brand partly because he is an avid dog lover (he is known for taking dog pictures from his wallet to show off when given an opening). Clow's agency initially took an assignment with Pedigree's parent company, Masterfoods (a division of Mars, Inc.), to work on a cat food brand. But all the while, Clow had his eye on Pedigree. He even remarked, in one of his early meetings with Masterfoods, "If I do a good job with cats, maybe you'll let me work on the dogs."

He got his chance in 2004. Pedigree was getting squeezed by premium dog food brands like Iams from above and private label offerings from below. To the outside world, and even, to some extent, within the company, Pedigree was seen as a midtier player and a basic, no-frills "dog food company"—a pretty low place to be on the food chain.

Clow felt the company could be more than that. He proposed a transformation that involved reimagining and then redesigning the way Pedigree ran its whole business—and even the way it thought about itself. "We had to help Pedigree discover their own culture and behavior," Clow says. "That meant we had to help them figure out what they believe, as opposed to what they sell."

Clow's team did a number of deep dives, immersing itself both in the company's history and in the universe of dog owners. What emerged about Pedigree was that it had been a pioneer in championing the whole idea that packaged pet food could be a healthier alternative to feeding dogs table scraps. Pedigree also ran a world-renowned research center on dogs in the UK, and the brand was known as a favorite among dog breeders. Meanwhile, the empathic research done among dog owners revealed that a surprising number of them were just like Lee Clow—they were mad for dogs. They humanized their dogs and considered them full-fledged family members. In fact, Clow's research found that most people wouldn't sit for a family portrait unless the dog was in the picture.

To try to get a sense of the experience people currently had shopping for dog food, TBWA set up immersive environments or "planets" where everything revolved around dogs and dog food. Among other things, the researchers in these rooms were bombarded with images from dog food ads—with lots of dancing dogs, talking dogs, and, said one researcher, "dogs climbing the Himalayas." What tended to be missing were images of real dogs, acting the part. Clow also noticed that dog food companies didn't seem to recognize or acknowledge that dog owners were *serious* about dogs—they care about what happens to them, how they're treated in the world. Clow felt that if Pedigree were to stake out a position as a company that cared deeply about dogs and was willing to demonstrate that through actions, this could provide "a springboard into a whole new kind of behavior for the company," Clow says. "It could inform product development, packaging, and lots of Pedigree's actions beyond selling dog food."

6.4 DESIGNING A NEW DOGMA

Clow felt that if the company were going to undertake such dramatic change, it might help to have a manifesto that articulated a new direction and philosophy. His agency designed an elegant company handbook titled "Dogma." Lushly illustrated with dog photographs, it laid out the new Pedigree philosophy in poetic language and included a letter from Pedigree's "top dog," CEO Paul Michaels, telling employees that the company intended to live by the dog-centric principles in the book.

"We're for dogs," the Dogma book declared in the opening pages. "Some people are for the whales. Some are for the trees. We're for dogs. The big ones and the little ones. The guardians and the comedians. The pure breeds and the mutts."

The Dogma manifesto was presented to employees by Michaels as "a compass to guide your daily decisions, as a quick inspiration on a tough day, and as a handy way to show friends and family what your job is all about." The book featured a series of stark, emotional photo portraits of dogs of every breed and in between. These weren't the show dogs of past ads, but "real" dogs who'd been found in the local area, in shelters, even in the offices of Pedigree and TBWA (and, yes, Clow's beloved German shepherd made it into the book). These dog portraits were also hung on the walls throughout the company.

The book was just one of the ways the company used design to begin to shift the Pedigree culture. Clow had advised Pedigree that if it intended to be a company for dog lovers, it should "walk the walk" by implementing dog-friendly policies in its own workplace. The company began to encourage associates to bring their dogs to work—and as part of that effort, Clow's team helped redesign ID badges and business cards so that they featured

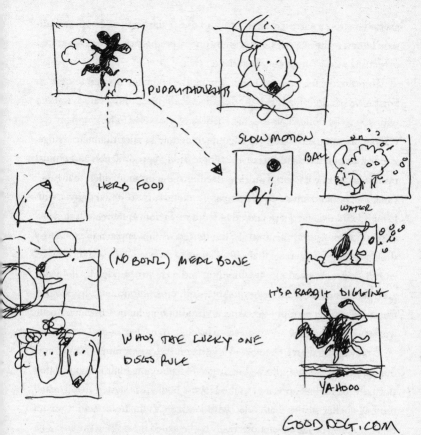

Lee Clow's early sketches for a "dog-centric" Pedigree ad campaign eventually led to a company handbook, featuring a Dog Bill of Rights.

images of their dogs. Pedigree also extended health care benefits to associates' dogs, opening the door for the company to become an advocate for other companies in other industries to do likewise.

The company took this all very seriously. In Japan, Pedigree cancelled a lease and moved out of its offices because the building wouldn't allow dogs. And Clow says that during one of the internal meetings, "we were going around the room sharing dog stories, and one of the Pedigree people admitted, 'I don't really like dogs, I was bit by one.' And this person actually ended up getting moved to another brand after that."

To really demonstrate its commitment, the company needed to take on what Mau would call a big, bold initiative. Pedigree found a cause, waiting patiently to be noticed, in the dog shelters of the world. The company developed an annual initiative to encourage people to adopt homeless dogs, and it was an immediate success. Pedigree didn't approach this as a typical corporate goodwill effort—it threw itself into the program such that it became a central element of the company's culture. In 2008, Pedigree, with Clow's help, designed a pop-up dog store in Times Square that featured a dog adoption center. Inspired by the design of the Apple Stores, Clow's design team tried to create that same feeling of immersion in a brand and its world. Visitors walked into the building and were surrounded by the Pedigree manifesto, the giant portrait shots, and, most importantly, live dogs—frolicking, having their photos taken, and auditioning for new owners. In one weekend, the Pedigree adoption store brought in twelve thousand visitors.

Gradually, Pedigree changed everything it did, in keeping with the new dog-centric philosophy. Packaging began to feature the same emotional dog portraiture that first appeared in the Dogma book. Ads started to tell heart-rending shelter stories that ended with Pedigree as the hero—and it wasn't hype, because Pedigree's actions really had rescued the dogs in the ads. The company Web site became a cyber dog park, where dog owners could upload photos and share stories. In the latest development, TBWA convinced Pedigree that it should declare and promote an international holiday for dogs, which made its debut on October 11, 2008. At long last, it really is true that every dog has its day.

SINCE THE transformation at Pedigree, the company has gained ground in all the standard measurement areas: annual sales, market share, ranking in its market. But what's particularly interesting is that Pedigree's food sales have spiked during each year's adoption drive. Clow points out that during the adoption drive period, the company hardly talks about its food at all; the focus of ads and other communications is entirely on helping dogs to find homes. Clow's conclusion: Good behavior does pay off, as long as it gets noticed.

And if the good corporate behavior is designed in such a way that it comes across as bold and interesting and relevant to people's lives, then it is more likely to be noticed—and talked about, and covered in the media. Mau preaches that if a brand designs its actions well, it can break through the media clutter in a way that advertising can't. The objective is to get people to notice what you're doing, instead of having to scream for attention in ads. Pedigree drew strong press coverage for its efforts, which is sometimes referred to as "earned media," as distinct from "paid media" such as advertising.

But perhaps the biggest payoff for the company has to do with the change in attitude among the people who work at Pedigree. "They used to come to work every day thinking they worked for a dog food company," Clow says. "Now they come in thinking they work for a company that loves dogs. That's a huge difference."

6.5 THE TURNING OF THE TIDE

Changing the culture of a midsize dog food brand is challenging enough. But what about an enormous and bureaucratic corporation that makes laundry detergent? Procter & Gamble has emerged as an interesting example of design transformation because P&G had to achieve this within one of the biggest, most process-driven and buttoned-down operations in the business world.

The transformation was led by A. G. Lafley, who took the company helm in 2000 at a time when P&G's stock was in the tank, though the company was still a market leader in many of its product categories. "Companies do this kind of change out of fear or out of pain," says Patrick Whitney of IIT's Institute of Design. P&G wasn't in too much pain, but "there was a lot of fear," Whitney says—in particular, fear of store brands from Wal-Mart and Target.

Lafley had two influences that drove him toward design: He'd spent several years working in Japan, "a very designed culture," he says, and (during his three decades with P&G) he'd also spent years doing in-home research. He believed in the power of up-close observation, and he'd noticed that P&G relied too much on what he called "research at a distance." Lafley felt that it was only when you ventured inside people's homes that you noticed funny little things that didn't show up in research surveys—like women having to resort to opening Tide packages with screwdrivers to avoid breaking fingernails.

Lafley knew that design firms such as IDEO and Continuum were on the cutting edge when it came to studying consumers up close and designing experiences for them. He came to believe that P&G could use design methods "to translate technology into products that delivered better experiences."

Once he'd had this epiphany, Lafley set out "to build design into the DNA of the company," he said in a recent interview conducted at Whitney's Institute of Design in Chicago. He started by hiring Claudia Kotchka as the company's new design chief in 2001. Together, they had to contend with a chemical engineering culture that was interested in hard, measurable results, not the fuzzy language of designing experiences. "They think deductively; they don't naturally think abductively," Lafley says of P&G engineers. And the company wasn't used to developing ideas in an open and collaborative manner or doing a lot of rapid prototyping and experimentation. When engineers were working on something, Lafley says, "they really didn't want anybody in the sandbox until they're ready to show it." He adds: "We nearly had a revolution on our hands."

But Lafley and Kotchka avoided an all-out showdown by starting slowly and choosing their battles. They felt that if they could successfully introduce the design process on small individual projects sprinkled throughout the company, they could begin to win over design skeptics. They went to divisions and products that were struggling and more receptive to change. Or, as Kotchka puts it, "We went where there was suction." Following upon the immense success of one of its early design experiments, the Swiffer (a mop that reinvented mopping via a swivel-head handle and disposable pads that absorbed dirt instead of pushing it around), P&G redesigned a couple of old standbys, Febreze fabric freshener and Oil of Olay (or Oil of Old Ladies, as the tired old product had been dubbed).

P&G hired outside designers and brought them into the company, but instead of setting up a separate design area, "we embedded the designers, and that was big," Kotchka says. "That's how you avoid the us-versus-them feeling." Not that the designers blended in altogether. "Designers are different," Kotchka says. "One of our first mistakes was to try to put them in cubicles," but the designers were having none of it. "So we changed the work spaces to be more open and collaborative. Some people in management then said, 'This looks cool, but what if everybody wants it?' And I said, 'That would be good.'" Everybody did want it, and P&G began switching everyone over to an open work space. Similarly, the designers balked at using PCs, so the company got them Macs—and soon everyone else was using Macs, as well.

The company had design schools—including Whitney's Institute of Design and the Rotman School—come in to give crash courses, and it also "sent people to go work on some wicked problems with IDEO," Kotchka says. She was amused at the initial reaction of some of the "Proctoids" upon their first exposure to IDEO's experimental methods. They reported back to Kotchka: "These people have no process; we need to teach them the P&G process."

Long known for fiercely protecting its privacy, P&G began to embrace a more open business model in terms of sharing ideas with outsiders; Lafley launched a "Connect + Develop" program that opened the door for collaboration with independent product designers. And to spark more creativity among its own people, the company set up a number of innovation centers both inside and outside of corporate headquarters—offbeat spaces where P&G staffers could "thinker" during the early stages of developing new ideas. This led to reinventing old standbys such as laundry detergent by coming up with ideas like the Swash product, for people who wanted to clean their clothes without washing them. As Kotchka notes, product developers began to realize there were infinite possibilities out there—that they didn't necessarily have to choose between the usual options A, B, or C because, as she says, "designers are able to invent option D."

6.6 "A RABBI INSIDE THE CORPORATION"

In attempting to transform themselves, Pedigree and P&G redesigned everything from dog adoption programs to kitchen mops. But what both companies were really focused on changing was corporate culture. "There's no ten-step program" when it comes to redesigning a corporate culture, says Continuum design's Gianfranco Zaccai, who has worked on transformation design with Procter & Gamble and other companies. "It's more about reaching a critical mass point." Zaccai is referring specifically to the point at which most people within that company begin to understand and accept that there is a new philosophy or sensibility in place. In the cases of both Pedigree and P&G, the companies were transformed at the moment employees started to actually believe they were working at radically changed companies—which then encouraged those employees to change their own behaviors (whether by bringing the dog to work or by switching from a PC to a Mac).

Ideally, though, the change goes deeper than that—and the people working at the transformed company become more creative and innovative, too. Innovation from within is what most companies are seeking when they approach firms such as Continuum, IDEO, Ziba Design, or Bruce Mau Design to work on transformation design. There is no simple recipe for innovation, of course. There are design principles and methods that, if applied well, can help. "We can instruct and train companies on the process of design thinking," says Sohrab Vossoughi, founder of Ziba. "But transformation must come from the top of the company."

P&G's Lafley is now held up in parts of the business world as a model in terms of showing how a nondesigner CEO can lead a design transformation within a company. Ziba's Vossoughi cites his client Ray Davis of Umpqua bank as another example. Even before knowing how he would go about redesigning the company, and before knowing some of the specific design techniques that helped create the "Umpqua experience," Davis was armed with a vision: He was determined to revolutionize the way the company interacted with its customers and committed to creating a more open, innovative corporate culture. Ziba helped provide some of the design tools needed, but Davis supplied the momentum for change.

For a company to embrace design, Milton Glaser once said, "there must be a rabbi inside the corporation," meaning someone who is not necessarily a designer him- or herself but is a true believer in the potential of design to reinvent old ways of doing things.

At the same time, however, the Institute of Design's Whitney says it's important to deeply embed design thinking and processes into a company so that they will endure after "the rabbi" leaves. Ultimately, the goal is to have everyone in the company begin to think like a designer: to question traditional practices and ways of doing things, to envision new possibilities, to be able to express and share those ideas, to collaborate within teams and begin turning the ideas into realities.

To encourage that kind of mass creativity throughout a company, the structural design of the organization is important, as is the design of the work space itself. The compartmentalization that Mau rails against is ingrained in the way many traditional companies are set up, both organizationally and physically—with departments separated from one another and individuals isolated in offices. There are good reasons why so many design studios, Mau's included, are set up as open lofts. The idea is to encourage as much collaboration as possible and to operate in a place where ideas intersect (increasing the possibility of smart recombination). Departmental borders, or silos, can become barriers; the design process works best when people can easily "jump fences" between departments and disciplines.

OXO is an example of a company that looks and runs like a design studio, even though it's a manufacturer. The New York City company uses wide-open work spaces and assembles multidisciplinary teams that work on projects from start to end (instead of handing off projects in stages, from one department to another). Rather than keep designers in a separate department, the idea is to "flow" design throughout the company. Desks are clustered together in horseshoe-shaped arrangements, with a large table in the middle so people can just informally turn around and meet when they need to. Sketches and prototypes are in full view and open to input from anyone ("And we're all very opinionated," says OXO's Michelle Sohn).

OXO's openness extends to the outside world. The company sorts through ideas submitted from outside inventors (it also tries to streamline that submission process so that paperwork and legal intimidation is minimized) as it looks for the occasional outside idea that fits with the OXO ethos. As Sohn tells it, that idea might come from anywhere. One came recently from an American minister who'd taken a trip to Africa and eaten a mango for the first time. "He couldn't believe how difficult it was to get into the fruit and how delicious it was once you did," Sohn says. "He was so motivated he sat down with a friend who's an engineer and they invented a mango splitter—and it's one of our bestsellers now."

Mau thinks it's critical that companies open themselves up to ideas coming from all directions, while also becoming more collaborative in their work processes. "That's why in our studio we insist on putting work on the walls, not leaving it on the computer," he says. Encouraging a more open participatory process not only increases the possibility of serendipitous connections, it also taps into the full body of knowledge within a company. It's all too common, Mau has found, for companies to expend great energy trying to solve a problem, only to have someone within the company remark afterwards, "I wish someone had asked *me*—I know all about that." You could be so much more efficient, Mau tells his clients, "if only your company knew what your company knows."

BECAUSE THE whole concept of "openness" is so central to design—designers must be open to new possibilities, open to a wide range of influences, open to constant feedback—a company that wishes to adopt a design-driven business approach must "open up" in ways that go beyond architectural floor plans. Becoming more open to risk and experimentation is particularly important, and can require a shift in thinking. Many executives were trained in business school to limit risk as much as possible, says Whitney. And that risk-averse attitude can permeate a business culture.

To counter this, companies must create what the Rotman School's Heather Fraser calls "a culture of courage" (IDEO uses the term "culture of optimism") in which risk-taking is not only tolerated, but encouraged. Fraser says that, in such an environment, a company and its leaders consistently exhibit "receptiveness to new ideas (good and bad) and an interest in every new insight." Experiments should be going on in plain sight and should be celebrated by the company, regardless of outcome.

Google's Irene Au characterizes that company's workplace as "an ecosystem of ideas where we just build stuff and see if it works." People get rewarded and promoted at Google "based on what they build and how well it works," she says. "And if someone, from any level or anywhere in the company, feels strongly that a certain thing should be built, you can make it happen here—even if the people at the top of the company don't necessarily think it's a great idea." Google's celebrated "20 percent time" policy gives engineers at the company one day a week to work on projects they are particularly passionate about (but at the same time, Google carefully measures and tracks experimental projects from the outset, and the company doesn't hesitate to pull the plug if interest in a new idea is slow to build).

In a market recession, there's a particularly strong tendency among businesses to limit risk by pulling back on creative experimentation. But lean times can actually be the best time to innovate—because bold and distinctive offerings will tend to stand out even more in a market where others are playing it safe. Experimenting in tough times is "like steering into a skid—it's counterintuitive" yet very effective, observes Marty Cooke, a creative director at the marketing and design firm Shephardson Stern + Kaminsky.

And it's important not only in terms of getting through present difficulties, but also for creating future opportunities. The iPod was initially developed during a down market, but later rode to success in an up market. Google CEO Eric Schmidt says that, in a recession, "if you tighten up too much, you eliminate future innovation and then you set yourself up for a really bad outcome five or ten years from now." Which is why, Schmidt told the *New York Times* as the financial market was reeling in late 2008, "we are going to continue to invest in small teams to do wacky things."

Whether in good times or bad, this commitment to exploration and experimentation requires a certain level of faith. "Instead of just moving in one direction and optimizing toward that," Whitney says, "companies must be willing to explore many options through prototypes—fully realizing that nine out of ten experimental prototypes might not be implemented." But Whitney also points out that when companies follow a design methodology that relies on up-front empathy and prototyping, they're actually *lessening* the chance that they'll go all the way out to market with something that then flops. "Executives hate risk, especially in tough times," Whitney says, "and yet they really can't afford to stop innovating. So what we're trying to show them is that design can be viewed as a way to 'de-risk' innovation—a way to make innovation a little less scary."

Which is not to say there isn't something slightly unsettling about design thinking itself—especially the *thinking* part. In some ways, it goes against the flow of the "just get it done, now" corporate culture, notes Jim Hackett, CEO of Steelcase, the office furniture company. In business today, Hackett observes, "there is an overcelebration of getting things done" and not enough patience for "thinking as part of doing." Hackett notes that the various stages of design thinking—the deep questioning, the attempts to frame complex problems, to consider many viewpoints, to "thinker" and make models—are all very contemplative in nature; he refers to it as "deep thinking."

A company that wants its employees to think more must give them the time, space, freedom, and tools to do so. In Hackett's case, at Steelcase, he provides on-site classes on "critical thinking" (which is really design thinking with a different label that's more approachable, Hackett says). There are many other pro-thinking policies that can be adopted: Mau points to the importance of scheduling workflow so that more time is built in to allow for idea generation and iterative testing of those ideas. And the Stanford d.school's George Kembel says that company executives and managers must, themselves, show an interest in how their people are thinking through projects at various stages, instead of just wanting to see the finished results.

EFFORTS TO design more productive and innovative companies are primarily intended to yield superior products and greater profit, but one side benefit could be the creation of better corporate citizens. Skeptics might take issue with that, particularly at a time when reckless and selfish behavior by companies has contributed greatly to the current problems we face. But design thinking—at least in the way it's supposed to work—encourages practitioners to step back and take a broader view of how one's actions affect the greater world. When this mind-set is adopted by companies, encouraging them to try to *Design what they do*, they're apt to do more constructive things—not because they're saints, but because of a raised awareness that all of those actions have a direct impact on overall success.

For designers, "integrity," from a structural standpoint, can make the difference between a chair that holds up and one that collapses. And in the same way, a company's overall integrity may become a much greater determinant, in years ahead, of its stability and long-term success. The public will be watching closely, looking for signs of corporate integrity or the lack of it. A study by Euro RSCG found that 70 percent of consumers feel business bears responsibility for driving positive social change. In a separate study, by Condé Nast, respondents said a company's overall behavior could and should be judged by four essential characteristics: transparency, commitment to quality, consistency, and authenticity. In short, people now seem to want companies to walk the walk, not just talk the talk.

Designers have taken note of these shifting attitudes, and many now see an opportunity to play a greater role in changing the way companies operate and behave. The IDEO designer Valerie Casey says she thinks there is a kind of "perfect storm" happening now—the swirling together of heightened environmental concerns and rising social activism, combined with technological empowerment of consumers—that is exerting new pressure on business to, in her words, "do the right thing." What's happening, according to Microsoft research director Bill Buxton, is that "the selfish interests and the ethical interests are coming together." As they do, it creates a need for synthesis—which happens to play to the strength of designers.

Design can weave together those selfish and ethical interests in a seamless manner, connecting business and social concerns. In reality, those interests have always been interrelated, even if companies didn't want to own up to that truth. As Mau notes, the old "unincorporated" way of thinking that held sway in business for so long—in which companies could behave as they pleased, as long as it was behind closed doors—contributed to the legacy of social problems that the designer in all of us must now begin to confront.

SOCIAL

7. FACE CONSEQUENCES

Coming to terms with the responsibility to design well and recognizing what will happen if we don't

7.1 DESIGN FOR THE OTHER 90 PERCENT

In the spring of 2008, about five months before the U.S. economy would crater in the fall, the floor of the Javits Convention Center in New York was awash in coffee tables and light fixtures priced well into the five-figure range. The event was the International Contemporary Furniture Fair, which tends to be a showcase for designed items that are over-the-top luxurious or amusingly quirky and stylized. Here, for example, you could find an $11,000 rocking chair for two that manages to complicate the act of sitting—you must work to maintain a delicate balance with a seating partner, lest you both tumble to the ground. If it's true that design can make things easier for those who have it hard, the ICFF demonstrates that design can also succeed by making things a little harder for those who have it easy.

Glitzy shows like this are the last place one would expect to find Architecture for Humanity founder Cameron Sinclair, yet there he was in the late afternoon, drawing a crowd away from the glittering booths to a far corner of the convention floor, even though the only chairs in this area were of the plain and functional folding variety. Sinclair—thirty-five-ish, blond haired, and boyish looking, who speaks rapid-fire with a slight British accent as he races through PowerPoint slides—was in full rant, sounding like a revolutionary who'd slipped inside the palace gates and was trying to stir things up. Sinclair's nonprofit group, AFH, is known for seeking out design opportunities in areas ravaged by natural disasters or man-made ones. The group's designers go in headfirst while others are fleeing, and they begin to study, draw, and build, creating everything from replacement homes to much-needed medical facilities to the occasional soccer field. "Someone once called us 'al-Qaeda for good,'" Sinclair told the crowd, "because we have all these sleeper cells of designers and when a disaster happens, they wake up and move into action."

His message to the young designers in attendance was: Forget these chandeliers and overpriced sofas; forget the whole New York design scene. Look to Africa and India, to the places where design is a matter of life or death. "Design like you give a damn," Sinclair said, offering up what has become his signature phrase.

The crowd ate it up. You got the sense they'd follow Sinclair anywhere, ready to help him make huts from sticks and mud, willing to deal with unconventional design challenges such as, *How do you make the building so the giant flying foxes can't get in?* (a real problem Sinclair once had to deal with while designing in Indonesia). One of the people at the furniture show, who'd gotten up and spoken right before Sinclair, was a woman in her twenties named Emily Pilloton, who'd been inspired, partly by Sinclair, to start her own nonprofit design group, Project H, which is trying to help distribute various ingeniously designed devices that can purify and help transport water in poor countries. "When I came out of design school, I was making chairs that sold for $2,500, and it just felt weird to me," Pilloton says. She saw what Sinclair and others were doing, and wanted to become part of that movement.

Sinclair is a good pied piper for the cause: He's a bona fide design rock star—sought after by governments, once funded by Oprah, repped by Hollywood's Creative Artists Agency. He is hot and so, too, is design activism in general these days. But there are good reasons why do-gooder design is "in" and why Sinclair is in demand, and they have to do less with trendiness than with the state of the world: It seems to be coming apart at the seams and crying out for anyone who knows how to mend.

The problems are so big, so pressing, that they follow Sinclair wherever he goes. As he told the audience at ICFF, on that very same day he'd come to New York to give this presentation, he was corralled by people at the United Nations, telling him they needed his help dealing with the crisis of the moment, an earthquake in China. "Somebody there at the UN said to me, 'By the way, there's five million people in China that need your help,'" Sinclair told the crowd. "And the thing is, I have only seven people working in my office. When you've got five million people that need help and you're coming to a guy who has less than ten people in his office—things are fucked!"

OR TO look at it a different way: There's never been a better time to be a designer, because there is so much in need of better design. "When things aren't working the way they should be," Bruce Mau says, "you have the makings of a great design project." And the bigger the problem, the more it challenges designers to question and rethink, to go deep in the investigation of the problem, to come up with original ideas and smart recombinations, to draw and build those ideas in order to make new possibilities visible and tangible.

As designers have increasingly applied these principles and this method with good results in the business realm, Mau and other designers at firms like IDEO and Continuum began to wonder: Could these same approaches be used to improve the ways people receive medical care or social services, the conditions of their housing, or the delivery of food and clean water? If designers have begun to figure out how to design "better experiences" for high-end consumers, what about improving the experiences of those who belong to the population segment that design activists have dubbed "the other 90 percent"? What about bringing design to the world that exists beyond the Apple Store?

Mau, whose roots in social design date back twenty-five years to his first efforts to start a do-gooder design shop, has been pondering and asking those kinds of questions for a while, but he didn't always feel as if anyone was listening. That started to change in the early 2000s, when he observed, while in the midst of putting together his Massive Change show, that a global movement was bubbling up in which people were attempting to apply design solutions to various social problems and challenges.

It's a good fit: Many social challenges are complex and require problem-solving that involves creativity, experimentation, empathy, and system thinking—all hallmarks of the design approach. Maria Blair of the Rockefeller Foundation, which provides grants to fund efforts in the social sector, says that her group has come to realize that "design has taken user-driven innovation to a whole other level" in the business world. "And so for us the question is whether that same process is applicable on social issues," she says. Blair notes that the people and organizations working in the social sector have always tended to be empathetic and "people-centered" in their approach to solving problems, but they've often lacked the ability to connect that understanding of a human problem with innovative solutions. That's why groups like the Rockefeller Foundation are now reaching out to IDEO and other leading design firms, in hopes of bringing fresh thinking to some old challenges.

For the design industry, meanwhile, it's a chance to perform on a bigger stage, with higher stakes (if not as much compensation). And, as Blair suggests, it may help design thinking move to a higher level. "If design is about how you solve problems," she says, "well, the social sector is about problems that we have not been able to solve. Those problems are incredibly complicated and they are people-centered. If the design industry can begin to grapple with them, it can move the whole practice of design thinking forward. It's one thing to design a better potato peeler—but addressing the issue of, say, clean water, is going to require another evolution in the practice of design."

What's also driving designers toward the social sector is a desire to bring more meaning to their work. Designers are known for striving to "make things better," which yields increasingly impressive devices and objects. But those "better things" don't always improve the world in the larger sense; in fact, they may actually make things worse by fueling overconsumption. Massimo Vignelli has said that he feels a lot of designers are caught up in the cycle of making shiny new objects that get thrown away when next year's model arrives. "If this is what you do," Vignelli says, "then you are feeding the junkyard."

To the extent that design has fed that junkyard over the years, then it could be said that the gravitation of designers toward the social sector is a kind of penance for past excesses—an attempt to "face the consequences" of their own success. But the guilt factor seems less prevalent than other, more positive impulses and motivations. A growing number of designers actually seem quite eager to confront global and social challenges because they view them as an ultimate test of creativity and resolve. As Al Gore noted in a speech at the TED conference in 2008, "How many generations in all of human history have had the opportunity to rise to a challenge that is worthy of our best efforts?" Such a challenge, Gore suggested, is one that should be embraced "with a sense of profound joy and gratitude."

It is also being undertaken with a sense of urgency. With a global population climbing toward seven billion, with economic turmoil, with resources dwindling and the planet warming, and with the spate of natural disasters that has hit hard in recent years, people have seen that governmental leadership has in the past not risen to these challenges, notes the designer Edwin Schlossberg of ESI Design. "All of this contributed to a feeling that began to build a few years ago and keeps growing," says Schlossberg. "The feeling is, 'I better do something.'"

CAMERON SINCLAIR first got that feeling about ten years ago: before Katrina, before the Asian tsunami, before 9/11. Each of those events subsequently added momentum to his quest and his fledgling operation, but Sinclair was first moved to action by the events in Kosovo in the late 1990s. At the time, Sinclair was in his late twenties, a self-described "CAD monkey," meaning he did computer-assisted design on a screen, working on retail spaces in New York. It was a living, but it didn't satisfy some of the romantic notions about architecture and design that Sinclair brought with him from the UK.

"I grew up in a rough neighborhood in south London," Sinclair recalls, "and I didn't know about the cathedrals and monuments—the world I was living in was concrete block towers." But Sinclair says that as he wandered the city he was fascinated by the way different architecture and design seemed to completely change the whole feeling of a neighborhood. "You could go into two neighborhoods that had roughly the same economic standing, but the structure of the buildings and the design of the neighborhood made all the difference—it could make a place feel safer, more vibrant. Because that environment was built with thought and understanding of the needs of those people."

When Sinclair went to architecture school in London in the mid-1990s, his notions about the social ramifications and responsibilities of architecture were somewhat out of step with the times. "What was being taught was the architecture of grandeur," he says. "It was about 'form follows finance'—or maybe, 'form follows fevered ego.' In any case, I was a black sheep there."

He remained restless during the early part of his architecture career in America. Then, as he was following news of the Kosovo conflict and the refugees there in need of shelter, Sinclair decided to get involved. He launched a competition to design transitional shelters for refugees. He expected a few people to submit ideas. More than two hundred design teams responded, proffering ideas to make housing in all kinds of low-cost, experimental ways—from rubble, or from hemp.

Sinclair used funding from the contest to begin building transitional housing in Kosovo, incorporating some of the winning designs. He then built upon that success by launching another design competition, this time aimed at addressing housing shortages in sub-Saharan Africa.

He learned some valuable lessons in Africa—one being that designers must pay attention to the hierarchy of people's needs. "We went to Africa to build housing," Sinclair says, "and realized the much more pressing problem was the growing pandemic of AIDS." And one of the biggest problems associated with that issue was that people couldn't get to hospitals—which meant that, somehow, the hospitals had to be designed to come to them. Sinclair's idea was to create mobile AIDS medical facilities, and again, he brought in ideas and raised support by hosting a design competition, this time to build the most innovative clinics.

Sinclair saw, firsthand, the power of design sketches to "make hope visible." When he showed some of the drawn plans for mobile health clinics to local doctors, their eyes lit up. "It gave them something to work with," he says. "Obviously, they knew all the issues involved, but without these plans they had no way of going to the head of the local medical organization or grant-making organization and saying, 'This is what we need, here's what it looks like, and this is how we would build it." Sinclair notes that in troubled areas, the role of designers is critical in part because they give form to possible solutions—"and the reality is, people don't fund problems, they fund solutions."

Getting from sketches to the finished design could be an adventure. As AFH began to move into unfamiliar territory, the architects and designers had to learn to adapt to unpredictable working conditions, scarce resources, and local idiosyncrasies. The aforementioned flying fox was nothing compared to the squatting elephants. In Sri Lanka, Sinclair's designers were building housing in an area that had begun to overlap with the migration routes of the local elephants, "and nobody told the elephants," Sinclair says. "So when they were walking through villages and needed a rest, they'd sit on someone's house." AFH had to work with a migration expert to figure out how to site the houses so that they wouldn't interfere with the routes of the elephants. "You don't get taught that in Columbia's School of Architecture," Sinclair observes.

Each time the world took a body blow, Architecture for Humanity gained strength from it. When 9/11 hit, AFH's membership quadrupled. "I think a lot of architects and designers started to ask themselves, 'What am I doing with my life?'" Sinclair says. As offers to help started pouring in to his cell phone faster than he could answer, he embraced an open-source model, giving anyone permission to start a local AFH chapter—and forty of them opened. Sinclair soon found he had upward of three thousand project designs on his laptop, coming to him from designers around the world.

The real turning point for AFH came in late 2004 and the following fall of 2005, with the one-two punch of the Asian tsunami and Hurricane Katrina. The latter was a particularly strong wake-up call for designers, perhaps in part because the disaster highlighted design failures on multiple levels—starting with the collapse of poorly designed levees, ending with government and social service responses that seemed to lack any semblance of a coordinated plan. Almost as soon as the floodwaters receded, a legion of designers, Sinclair among them, rushed in to face the storm's consequences and, in the period that followed, the Gulf Coast basin became a design petri dish.

7.2 RISING FROM THE WRECKAGE

The Katrina debris, in some cases, formed the raw material for designers. For example, one volunteer design group had the idea to actually sift through wreckage and use salvageable materials to construct new furniture. The group, which called itself the Katrina Furniture Project, also began training Gulf Coast residents to make tables and chairs from destroyed building stock.

But the furniture wasn't of much use to those who lacked a home. In the year after the storm, designers and architects came to the Gulf Coast to help rebuild lost homes. One of them was Marianne Cusato, a young designer based in Florida who traveled to Biloxi, Mississippi, in the fall of 2005. Cusato learned that the most pressing need was for an alternative to the FEMA trailer, the standard temporary housing unit provided by the Federal Emergency Management Agency. The trailers were, in Cusato's eyes, not much more than "tin cans"—small, cramped, dark, unsightly, and uncomfortable. And worst of all, Cusato says, they apparently were meant to be that way.

"There had been no attempt to make the trailers comfortable," Cusato says, "because the rationale was that if you did that, people would just stay in them instead of looking for a home." In fact, at the time of Katrina, there were still people in the region living in FEMA trailers left over from Hurricane Andrew, a decade earlier. To Cusato, this was an argument for making the temporary housing better and more livable, because, she says, "the reality is, what we put in place as temporary housing will, in many cases, become permanent. So my feeling was, 'Let's be honest about this—some people are going to stay in the housing you provide.'"

Cusato felt the real question was, *Can you design temporary housing that is decent and dignified enough to work in the long term?* "Our pitch to FEMA was, 'We can build something for the same low price, and here's the best part—it's something you won't have to dispose of afterwards.'" (She knew this would be appealing because it can be troublesome and expensive for the government to get rid of old abandoned trailers.)

She started with the idea of building a "cottage" instead of a trailer—not just for aesthetic reasons, but for practical ones, too. A cottage lent itself to multiple uses; if you built it on the back part of the property lot, "then, if and when the person rebuilds their full house, the cottage becomes an asset—a guest cottage, a rental unit, a mother-in-law flat." Her other idea was to design the cottages so that they could grow. "So the cottage itself becomes the first piece of the larger house," Cusato says. "Over time, as your insurance money comes in, you build extensions."

In terms of the design of the cottage itself, Cusato knew that to make it as affordable as possible, she would have to streamline and simplify—which, ultimately, became part of the cottage's appeal and charm. Merging the style of an English seventeenth-century cottage with touches of Southern architecture (she was also influenced by the simple and adaptable Sears, Roebuck mail-order cottages of the pre-WWII era), the cottage was just three hundred square feet, but Cusato made it seem bigger with high nine-foot ceilings and tall vertical windows. Inside, she included crafted details like built-in bookcases to make the rooms homier.

The pitched metal roof and cement board siding were built to be both storm resistant and energy efficient. One of the most important features of the cottage was the large attached porch: Cusato reasoned that its spaciousness would have the effect of making people feel less restricted while also encouraging a sense of community with other cottage owners. Even with the added features, Cusato managed to bring the project in at a cost of $35,000—

which was achieved by making sure everything in the home had a purpose or, better yet, multiple purposes. "Anything gratuitous was stripped away," she says.

The Katrina Cottage, as it came to be known, impressed government officials enough that the U.S. Senate approved a billion dollars to build cottages throughout the Gulf Coast disaster areas. And Cusato's cottage earned the 2006 People's Design Award from the Smithsonian Institution (beating out Apple's iPod, no less). But the real validation for Cusato came when she brought a model version of the cottage to the annual national home builders show. "We set ourselves up alongside these big, elaborate show homes," Cusato says. "Right next to us was a 'Next Gen' house—the kind where you can turn on your stove from your cell phone and where there's a TV in every bathroom." But people at the show seemed more interested in Cusato's little cottage than in the fancy houses.

Then Cusato started getting offers from builders, who thought her design would be perfect for lakefront cottages. Cusato was taken aback. "Initially, we thought, we don't want this to become all about catering to rich people, which so much architecture does. And then we figured, you know, if you have a situation where disaster housing is exactly the same as the housing that rich people choose to vacation in—well, that's a good thing, isn't it?"

WHILE CUSATO was working on her cottages, Cameron Sinclair was also on the ground in Biloxi. Sinclair was struck by the lack of progress there in actually getting houses built. Government agencies were providing loans, but there was no coordination in terms of helping people work with contractors to get housing built. One of the golden rules of social sector design is that you have to collaborate with local groups, so Architecture for Humanity immediately formed partnerships with local relief workers and established a community board to help figure out who needed housing help the most. Sinclair also reached out to his growing network of designers and architects and invited a number of them to come down and join him in creating an array of model home designs. One of the two dozen or so who accepted the challenge was Marlon Blackwell, an architect based in Fayetteville, Arkansas. "I'm not a joiner usually," Blackwell says. But Katrina was different: "Everyone in my office just said, 'We gotta do this.'"

If his principle of *going deep* is essential in designing for business applications, it may be even more critical when trying to design solutions in the social sector. Blackwell and the other designers who went down to Biloxi as part of the AFH mission felt that before they built anything, they had to spend time with the storm's victims. Designers walked through disaster sites, visited people living in temporary shelters and trailers, and began to sketch out ideas based on what they saw and heard. Blackwell had an idea that got into his head and wouldn't leave. It had to do with a dog.

In his younger days, Blackwell had put himself through school selling Bibles door to door in the rural South and became intimately acquainted with what's known as a "porch dog"—which, as he defines it, is "that old hound dog that lays out on the porch and greets people when they come up, but also is the fierce defender of the house." The metaphor fit with what Blackwell felt was needed in new houses in the Gulf Coast: It had to be a house that could guard and defend against the next storm, but could also be welcoming, "a house that turns its underbelly to the public."

The plans for Blackwell's Porchdog house called for metal skin, in the form of sliding panels and louvers that could roll into place when it was time for the house to hunker down. The panels could be slid into place by hand—the first thing you lose in a storm is electricity—and the louvers could also be opened to let a breeze blow through the house. The welcoming "underbelly" of the house was the porch, which became the interface with the street, while the rest of the house was raised up to protect from flooding.

Blackwell felt the emphasis on the porch connected the house to the traditions of Southern culture, and to the neighbors. But he acknowledges that at the same time, the house looks anything but traditional—which puts him at odds with "new urbanists" in the area (Cusato among them) who felt it was important to preserve traditional styles in the rebuilding process. To Blackwell, the postcatastrophe situation was a time for Darwinian evolution in design. "I felt this wasn't a time to recede into the comfort of what's known," he says. "It's a time to generate new building types that can address the volatile relationship of living between water and land."

He wondered how the Porchdog would go over with local residents, and he found out when AFH's architects showed their plans to the community one day at a gathering held in a Salvation Army hut. The idea was that each architect and his/her design would be matched up with one family

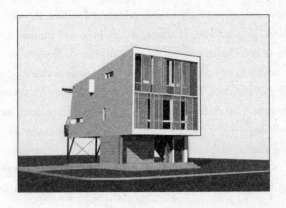

in need. Sinclair and the other organizers added a nice touch by allowing the storm victims to be the decision makers—they were given a chance to stroll past the designers who were, in effect, "auditioning" for them with drawings and models. In the end, one family would choose one architect, who would then become their personal home builder. Blackwell describes it as being like "an architectural flea market," but also says it was a delightful experience.

A few of the people strolling past looked at the model of the Porchdog "and just shook their heads," says Blackwell, laughing. But one who took a closer look was Richard Tyler, a single father of two who'd lost his house in the storm and was living in a FEMA trailer. Tyler was accompanied by his fourteen-year-old son, who was taken with the design. Blackwell joked that the Porchdog house came with a dog. Tyler said they didn't need a dog; there were plenty of those available after the storm. But they did need a house and they liked this one.

Blackwell spent the next two years building the challenging Porchdog. While he was working on the house, he asked Tyler if he wanted to make any changes to the design or customize it in any way. "He told us, 'Just do what you think is best,'" Blackwell says. "At one point he nodded over to the FEMA trailer and said, 'Look where I'm living now. Anything you do will be an improvement in my life.' "

BLACKWELL'S HOUSE was one of the last to be completed in Sinclair's Biloxi Model Home program. "People talk about the Gulf Coast like it's our great shame," Sinclair says, "but the real story is how people went down there and rebuilt houses, one at a time. You don't see that very much on the news, because houses being built doesn't make the news."

While Blackwell's Porchdog and Cusato's Katrina Cottage are markedly different, in each case the designer found that within disaster was an opportunity to reinvent and to have a life-changing impact. Sinclair points out that most designers hunger for those kinds of opportunities and don't get them. In the business world, "designers can go a long time without ever seeing their work realized," he says. "I think the current generation of designers is having a midtwenties crisis. You've just done seven years of education and honed your skills so that you're on the cutting edge of technology, materials, construction. You're in a position to do so much. And then you go out into the workforce and you're basically a CAD monkey, doing nothing of any significance. You get to about twenty-eight, twenty-nine, and then you freak out."

7.3 HOW TO MAKE SPAGHETTI IN 300 EASY STEPS

Some aren't waiting until they get to the ripe old age of twenty-eight. At the leading design schools, a culture shift has taken place in the last five years or so. These days, a design activist like Sinclair is far more likely to be seen as a role model than is a fashionable products designer like Karim Rashid.

John Bielenberg, who leads a design student workshop called Project M, notes that "in the 1980s and '90s, design students were more career motivated—it was about getting a job and your Saab convertible, and achieving fame in the design world." In the current decade, Bielenberg says, students have begun to think about the potential to apply design skills to larger issues and problems. "I don't think it's so much about 'We're going to save the world,'" he says. "It's more a matter of just wanting to use your talents for things you can feel passionate about."

This is evident these days at the Rhode Island School of Design, which, in the past, tended to be more associated with the art side of design. That hasn't entirely changed. At RISD, located in the small, bustling city of Providence, the campus buildings are spaced well apart from one another, and on any given day you see students making the trek through town as they lug sculptures and various unrecognizable objets d'art.

But some momentum has shifted from making beautiful objects to tackling messy problems. For example, RISD set up a kitchen that was monitored with cameras, so that people could be observed cooking meals. The idea was to see how many physical steps were involved in the process, and how much bending and reaching was necessary, so that the school's designers could begin to think about how to redesign kitchen layout and appliances for those who are older or physically impaired. Former RISD president Roger Mandle tells how everyone was shocked to discover that preparing a simple spaghetti dinner required the cook to take about three hundred separate actions or steps.

It's not just the kitchens in Providence they're trying to fix at RISD. In one classroom where an end-of-semester "crit" was under way (in which student design projects are reviewed by professors), a couple of twenty-year-old students showed sketches and prototypes for a school dining hall they hoped to build in Tanzania. The project came about because one of the students, Laura Sussman, had gone to the area on a volunteer mission and was told by a local middle school official that kids in the village of Pommern had no decent place to eat. Sussman took the idea back to RISD and was given the green light to pursue it as a student project.

In the RISD critique session, as she and partner Elliott Olson explained their blueprints, the attention to detail was impressive. Using corrugated tin walls and local materials that could be easily gathered, with a construction model simple enough to be built by amateurs, the structure made use of movable walls to provide maximum versatility, had a warped roof that collected rainwater and poured it into containers, and paid strict attention to matters of ventilation (in the old kitchen that this was replacing, Sussman mentioned, "people had actually died" from excessive heat and lack of air). The whole thing was designed to be built for somewhere in the neighborhood of $15,000, which impressed Sussman's professor, Liliane Wong. "It takes advantage of every single economy possible," she commented, which made it sound like an A grade might be in the offing.

Sussman, who was about to graduate, wasn't sure what would happen next for her and for the project but said she really longed to go back to Tanzania. "It's easy to see a career full of making office buildings or high-end restaurants as a natural path," she says. "But hopefully the Tanzania project will be a start to a career in work that actually matters." Sussman mentioned the "human camaraderie" involved in such projects, but she said the overriding appeal can be boiled down to a basic sense of responsibility: "The idea is simple," she says. "If you can help, do it."

Her professor seemed to have a similar outlook. Wong had checked out of a promising career with a private architecture firm "where I was losing my soul," she says. At one point, she was volunteering in a soup kitchen and had her glimmer moment: She saw that even a minor detail—such as whether the soup kitchen's carrots were chopped in a way that made them chewable for homeless people with dental issues—could have an impact on people in need. She went back to RISD and suggested that students begin to view the daily lives of the homeless as a design issue. This led to a fascinating set of student furniture designs, all geared to the particular and idiosyncratic needs of people in transitional housing.

The students dove deep into the ethnographic research, spending time in transitional housing, watching, trying to get a sense of what was and wasn't working for the people in their daily lives. In many ways, the students were doing exactly what IDEO does for Fortune 500 companies, sans the research budget. Based on the observations, the students designed special slotted medicine cabinets for mental health patients who had to contend with a be-wildering array of pills. They built a linear shelving unit for two men whose prized possession was a collection of LPs from the 1980s. There was a lounge chair made for someone who'd spent years sitting hunched on the street.

One of the most intriguing pieces was a dining table with a detachable tray on wheels (facing page), designed to serve not just a real need but also an imagined one. The woman for whom it was built eats alone in her room every day; hence the tray section. But when that smaller section is inserted back into the end of the main table, the table expands to seat six—to accom-modate the dinner party that this woman dreams of having. It's not likely that the dinner party will happen ("She doesn't know anyone," Wong con-fides), but it's important for her to believe that it will. "In some cases, we're designing pieces about hope," Wong says.

What's happening at RISD is going on at many other design schools around the world. Just to single out one example, students at the University of Minnesota came up with a particularly clever design solution for a problem

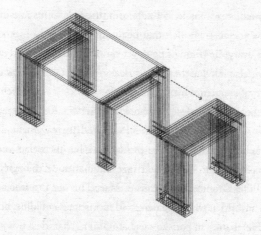

that was causing great angst among local homeless people. The students learned, after some deep-dive research, that when these people were chased by the police from their favorite squatting area under a local bridge, they often ended up losing all their possessions, which were either taken away from them or left behind in the confusion. The solution the students came up with: camouflage backpacks that are also magnetized. When the cops are approaching, you toss up your pack so that it sticks to the girders or the underbelly of the bridge, hidden there until you come back for it.

RISD president John Maeda has noticed this wave of design activism in schools, including his. "We're going back to an emphasis on, 'How do we use our talents and abilities to attack major social problems that we thought were unsolvable,'" Maeda says. "I think in design, in some sense, we're going back to the sixties."

ACTUALLY, WE may be going back a lot farther than that. The real heyday for design activism was probably a century ago, when the Arts and Crafts Movement was transitioning into the Bauhaus and Modernism eras. A common thread of all these movements was the belief that design could improve life and create a better world—not just for some, but for all. In fact, one of the first "star" designers, William Morris, best known for his wall coverings and patterned fabrics made in England in the late 1800s and still popular today, was a socialist who felt that beautifully designed objects should be available to everyone (though he never quite worked out the economics of that). Morris also felt that these finely designed creations should be made by individual craftsmen, not by machines.

When the Bauhaus design school emerged in the years after Morris died and his Arts and Crafts Movement subsided, there was a new acceptance by designers of mechanized mass production, but the belief in egalitarianism remained fervent. The new rising luminaries of the early twentieth century—Walter Gropius, Le Corbusier, Marcel Breuer, Buckminster Fuller—envisioned a world in which housing and furnishings could be well designed yet affordable, thanks, in part, to mechanization. That ideal was still holding strong midway through the century, when the husband-and-wife design team of Charles and Ray Eames had as their mission: "To get the most of the best to the greatest number of people for the least."

The design activists of the early twentieth century can be seen as egotists (after all, who were they to think they could make the world better?), though Bucky Fuller, for one, claimed to have no ego at all, having killed it off when he was in his early thirties. At that time, following the death of his daughter, a devastated Fuller was on the verge of suicide, but instead famously committed "egocide"—deciding to dedicate his life to using design to help others. It's reasonable to think Fuller could have earned more riches, and perhaps reaped more glory, by designing luxury houses or automobiles instead of domes and three-wheeled wonder cars. But the narrow ambitions of commercial design weren't enough for Fuller. By tackling life-on-earth challenges, he could ask deeper questions and try to jump higher fences. In this regard, Fuller can be seen as a prototype for today's social activist designers.

SOMEWHERE ALONG the line, the fix-the-world optimism that was so much a part of the design world in the first half of the twentieth century began to dissipate. Charles Eames had hoped that the growth of industry would help disseminate good design and better living; but Eames seemed disillusioned by industrial waste and pollution when he declared in the 1970s, "We've realized our dreams at the expense of Lake Michigan." By that time, industrial design had become associated with other negative developments, too, including planned obsolescence—"the shame of design," in the words of Brian Collins.

As for the ambitious public housing designs that came from some of the modernist architects, they had their own design flaws built in, albeit not intentionally. Housing projects and buildings designed for the masses came to be seen as faceless and impersonal; some were structured in ways that led to unexpected consequences, concluding in abandonment. "All those best-laid plans went seriously astray," wrote the *Time* magazine architecture critic Richard Lacayo, reflecting on past attempts at social design while reviewing Mau's 2004 Massive Change exhibit. Lacayo thought Mau's show echoed the old boundless optimism, and he reminded readers that some of those earlier design dreams didn't end well: The public housing that Le Corbusier and others envisioned as "tall buildings on sunlit green plazas," Lacayo wrote, "ended up as dirty towers on windswept lots, the kind of places we have been critiquing in recent years with dynamite."

By the time Cameron Sinclair's generation was coming on the scene, ambitious efforts to design for the social sector had been somewhat discredited. "I think the design world became gun-shy with regard to this ideal that design could change the world," Sinclair says. "People had looked at these utopian dreamers from earlier years and decided, 'These guys didn't know what they were talking about.'"

While architects were pulling back from more socially ambitious projects, the graphic design world—which had always been a bastion of strong political expression and social-issue advocacy, often expressed through posters—had become commercialized to the point that by the mid- to late twentieth century, some designers felt the industry had lost its soul. Milton Glaser expressed this identity crisis in a humorous way with his twelve steps on "The Designer's Road to Hell" (next page).

Glaser's 12 Steps on the Designer's Road to Hell

1. Designing a package to look bigger on the shelf.

2. Designing an ad for a slow, boring film to make it seem like a lighthearted comedy.

3. Designing a crest for a new vineyard to suggest that it has been in business for a long time.

4. Designing a jacket for a book whose sexual content you find personally repellent.

5. Designing a medal using steel from the World Trade Center to be sold as a profit-making souvenir of September 11.

6. Designing an advertising campaign for a company with a history of known discrimination in minority hiring.

7. Designing a package aimed at children for a cereal whose contents you know are low in nutritional value and high in sugar.

8. Designing a line of T-shirts for a manufacturer that employs child labor.

9. Designing a promotion for a diet product that you know doesn't work.

10. Designing an ad for a political candidate whose policies you believe would be harmful to the general public.

11. Designing a brochure piece for an SUV that flips over frequently in emergency conditions and is known to have killed 150 people.

12. Designing an ad for a product whose frequent use might cause the user's death.

7.4 ENTERING THE ERA OF "DISTRIBUTED POSSIBILITY"

That design activism was jump-started during the Bush years doesn't surprise observers such as Edwin Schlossberg; as he noted previously, inadequate leadership can spur inventive design because it motivates people to take matters into their own hands. Beyond that, the early 2000s were also, according to designer Jakob Trollbäck, "the iPod moment," a time when designers "emerged as pop stars and the primary drivers of consumption," and this made some within the design industry increasingly uncomfortable. Writing in his 2005 book, *In the Bubble*, John Thackara commented: "We're filling the world with stuff—but what value does it add to our lives?"

Something else was happening around this time, and Mau sees this as the key to the rise of the new design activism. While bad leadership, growing problems, and rising social concerns all provided the motivation, technology began to provide the means. We've seen the rise of what Mau and others have referred to as "distributed possibility"—meaning the widespread dissemination of design tools, useful knowledge, and expanded capabilities—all being downloaded and passed around as never before. Today, a solo designer can go online and learn about a problem, find out what's been tried and what hasn't, download technical data, connect with experts, and seek out collaborative partners. When you've got a finished product, or maybe even just a rough prototype, you can show it to the world on YouTube. "These are tools that in the past were only available to monarchs, and maybe movie producers," Mau says.

And what it means, according to Mau, is that more than ever before, change—including solutions to our problems—is going to come from the ground up, not from the top down. Forget about governments solving problems: "In government, there is almost no distributed possibility," Mau contends. "People within those bureaucracies don't have the power to do much of anything." They may not even believe, in their hearts, that the problems *can* be solved. But the naïve outsiders—all those basement Buckys, the ones willing to ask the stupid questions and build crazy models—are more empowered and emboldened to try to fill the void. "That combination of great tools and a great calling to solve problems," Mau says, "is what is going to make this a very interesting time to live in."

THAT KIND of grassroots-level design activism was on display at an event in a New York City gallery in the fall of 2008. Billed as Designism 3.0, it was the third iteration of a sort of collective outcry by designers anxious to make a difference. The movement was organized by Brian Collins, who turned to Milton Glaser for help (Glaser came up with the Designism name, and made a logo to go with it). Collins first started Designism about midway through the second term of the Bush administration (a period Glaser referred to at the time as "this mad dream we've been dreaming"). "It just seemed to me that with all that was going on, there was little dialogue about using design to make the world a better place," Collins says. In the years prior, "all you heard about were celebrity designers or the 'starchitect' of the month."

The first couple of meetings featured a lot of venting about Bush and some earnest soul-searching about the role of designers in society (all of which prompted the *Vanity Fair* columnist Michael Wolff, a guest at the second session, to accuse the designers of navel-gazing). But by the third session, Designism had started to move beyond words into action. Designism 3.0 had the same inventive, forward-looking feel that Mau's Massive Change show had. The hall was filled with two hundred mostly young designers, a number of whom came up to the front of the room and showed what they were working on at the time.

One of the projects on display was the Aquaduct bike, which was made by a team of five young designers in California to address water shortages in Africa. Here was the idea: To get more drinking water to people in areas where it's scarce, the local water must be a) purified and b) transported. The designers reasoned, why not do both jobs simultaneously with one device? They created a bike that when pedaled, generated energy for an onboard condenser/purifier. The cyclist could ride down to the nearest body of water, fill up a tank attached to the bike, and then head home, knowing that by the time he/she arrived, the water would be drinkable. The first version of the Aquaduct bike was made in a garage, with basically no budget (the designers worked for IDEO but did the project on their own so they could enter it in a contest sponsored by a bike company). In the past, this might have been one of those crazy ideas that never gets *beyond* someone's garage. But in the age of Mau's "distributed possibility," the prototype was seen on YouTube by thousands and the designers were even contacted at one point by Bill

Clinton's Clinton Global Initiative. The designers and various partners who've approached them are still trying to figure out how to mass-manufacture the Aquaduct affordably, so that it can be brought to the places that need it most.

The story behind the Aquaduct brings to mind what the journalist Thomas Friedman has said regarding what America needs to succeed in a dynamically changing world: "We need a hundred thousand Dean Kamens in garages." Well, not to worry—they're out there. One open innovation network, InnoCentive, has 160,000 "solvers" in its group, all of them looking for problems to tackle. Dwayne Spradlin, head of InnoCentive, says that, although there is prize money involved, the primary motivation for these people is "they want to work on problems that matter. And the harder the problem, the better. These people do it for the fight."

And these basement Buckys will go to great lengths to bring their ideas into the world. For instance, Steve Mykolyn, whose "15 Below" coat for the homeless was one of the featured items at the Designism 3.0 event, almost turned himself into a human Popsicle to prove that his design worked. Mykolyn, a creative director at a Toronto-based ad and design firm called Taxi, started with a smart recombination. He'd observed that professional bike racers stuff newspaper under their jerseys to stay warm. Mykolyn's idea: construct a coat for the homeless, designed so that it could easily be filled with insulating newspaper. He spent several months working on the concept with a fashion designer friend, using rainproof fabric lightweight enough that the coat could be folded to fit into a pocket or form a pillow when not being worn. The coat's lining was designed with a series of hidden Velcro pockets running along each seam and in other strategically placed areas—the pockets could be opened and stuffed with crumpled newspaper until the coat puffed up like a down-filled parka. The wrist and waist areas cinched closed for a tight seal.

Mykolyn wanted to put his prototype to the ultimate test, so he decided that he would hang out in a local commercial meat freezer wearing the newspaper-stuffed coat. In the days before he did his test, he heard that the men who worked at the freezer were betting with each other on how long he'd

last (the plant manager didn't think he'd make it past an hour). "When I got there, they had a full-blown pool going," Mykolyn told *Creativity* magazine. "It was kind of intimidating because these guys who work there wear arctic gear, two toques, mittens, boots, two giant heavy coats—and there I was with this polyester coat stuffed with newspaper."

He proceeded to spend eight hours in the freezer and still wasn't cold. So, for good measure, he ventured into a "blast" freezer, where they kept ice cream at temperatures that dipped down to minus 40 degrees Celsius. He says, "I spent forty-five boring minutes in there before I thought, 'You know what—the coat works. Let me out.'"

Using the Web and his contacts in the media, Mykolyn began to circulate word about the 15 Below coat. Before long he had celebrities like Jon Stewart and Yo-Yo Ma agreeing to sign coats that could then be auctioned off so that many more coats could be made. And Mykolyn teamed up with the Salvation Army, which handled the logistics of handing the coats out to homeless people in Toronto. "Originally, I thought maybe you could just walk around and hand out coats to people," Mykolyn muses. "But it's weird—sometimes when you try to give someone on the street a coat, they get really mad at you."

It's not that weird, actually: People can be offended by well-meaning attempts by outsiders to offer designed solutions (it's a fine line between helping and meddling). Then, too, people may not know what to do with those designed solutions once they're handed to them. Sometimes it's just extremely difficult to get it to them in the first place. Designing a better world gets complicated.

7.5 COMING TO TERMS WITH THE "YOU'LL ONLY MAKE THINGS WORSE" SYNDROME

While designed "things" can be very useful—Mykolyn's coat, for example, or the Aquaduct bike—they're usually just one piece in a puzzle. "One of the most important things designers have to learn if they're going to work in the social sector," says the Rockefeller Foundation's Blair, "is that success in this realm is not about designing the next 'thing.' It's about designing better systems." For example, just making a better water filter won't solve water problems, Blair points out. You must figure out how to distribute the filter, who pays for it, how to get people to use it, what to do in areas where there is no water to filter.

This means that designers working on social problems must, to use Mau's expression, "expand the problem" and look at it in a larger context. When you do that, the original problem you're trying to solve may, as Mau notes, turn out to be many interrelated problems, each of which must be addressed—which is what makes these problems so "wicked," from the standpoint of the would-be solver. This expansive approach tends to run counter to the more conventional method of trying to simplify problems, to boil them down in an attempt to come up with "the answer." Often, there is no *answer*, Mau maintains. However, there might be ten answers, or twenty-seven.

If a problem, and especially a wicked social problem, isn't looked at in this expansive way, then it's quite possible the designer may solve one aspect of the larger problem while making another part worse. The fact that this has happened in the past (e.g., the aforementioned public housing projects), and will surely happen again, forms the basis of the argument that designers should perhaps steer clear of complicated social problems altogether, because of the distinct possibility that their efforts will only make things worse.

To add weight to this argument, it is sometimes pointed out that designers have been known to try to solve problems without sufficient knowledge of the particular situation. This leads to the compelling two-part argument against design intervention: *You don't know what you're doing*, and *you'll only make things worse.*

Designers themselves have acknowledged concerns in this area. In a 2008 essay posted on the Design Observer blog, John Thackara, who has a long history of being involved in social sector design efforts, suggested that well-meaning itinerant designers "often lack in-depth knowledge of local ways of building and living, and propose solutions that cannot be readily adapted to local conditions and are therefore unlikely to be sustainable." Thackara wrote that he's come to the conclusion that designers should mostly stay on their home turf, because "we can usually do more good in our own backyards than in foreign parts." This prompted Architecture for Humanity's Sinclair to respond with a post of his own, writing: "Yes, there are dozens of 'examples' where we can point to designers screwing up, getting it wrong, undervaluing the input of the community. Yet there are hundreds of stories where quiet moments of innovation have been an element of incredible change in a community."

Whether designers should focus on solving problems close to home, as Thackara suggests, or go where the need for their services seems most urgent at any given time, the Sinclair approach, is debatable. But it's awfully hard, especially these days, to make the case for doing nothing—which would be about the only way designers could completely avoid making the occasional misstep that leads to unintended consequences. Better to accept that designers *will* make things worse at times, but they're far more likely, if they properly employ the best practices of the trade, to make things better.

As the Rockefeller Foundation's Blair notes, cutting-edge design thinking may represent one of the most promising tools currently available for taking on big, thorny problems.

Some of those problems might not be as bad as they are now if good design techniques and strategies had been employed earlier. Collapsing bridges, failed levees, breached airline security systems, compromised elections, social programs that don't function, and economic systems spiraling out of control: All of these crises can be at least partly attributed to a lack of rigorous design thinking, contends Thomas Fisher, who heads up the University of Minnesota design program. In most of these instances, Fisher says, it took a while for the design flaws to become apparent, but now we've reached stress levels—due to population growth, environmental degradation, and general wear-and-tear—that have brought us to the cracking point. Fisher uses the term "fracture critical" to describe systems and objects that were not built to withstand a single part failure. Hence, a weak gusset takes down a bridge, and one failed bank (Lehman Brothers) triggers a series of events that results in financial chaos.

Fisher contends that the use of rigorous design processes—"going deep" to more fully understand needs and situations; following an iterative process with prototypes, so that design flaws can be detected early and fixed; building in design qualities such as forgiveness and integrity; engaging in scenario planning, to try to envision various possible outcomes and consequences—is the key to rebuilding our infrastructures and social systems so that they won't fracture quite so easily in the future.

It doesn't mean designers won't still get it wrong sometimes—but good design can at least increase the ratio of intended versus unintended consequences. And, too, they can hope that the unintended consequences that do arise may occasionally turn out to be pleasant surprises. Emily Pilloton, the young idealist designer who appeared along with Cameron Sinclair at the furniture fair in New York, shares a story of one such surprise, involving her experiences in Africa. Pilloton's Project H group has been working to distribute a simple yet effective design device known as the Hippo Roller, which allows the user to transport water more easily by rolling it instead of carrying it.

According to Pilloton, in one small African village where the device has been widely adopted, the unintended consequences caught everyone off guard. "It turns out that literacy rates among women in the village have gone way up," Pilloton says. The reason: Women are spending about one-fifth as much time lugging water around, which frees up time for other things. And that's not the only change wrought by this simple plastic Hippo Roller. "Men have started fetching the water now," Pilloton says. "Before, they wouldn't carry water, but they think using this tool is more manly and fun. And I noticed they've started taking their sons with them—so it's becoming a father and son bonding thing, to go and get the water."

Pilloton never intended any of that, but she can live with those consequences.

8. EMBRACE CONSTRAINTS

Design that does "more with less" is needed
more than ever in today's world

8.1 LESSONS FROM THE "RAMBO SCHOOL OF DESIGN"

For much of his professional career, Jock Brandis served as a mister fix-it on Canadian movie sets. If a special camera or lighting rig was needed for a difficult shot, Brandis would cobble it together, using whatever materials were close at hand. He had to work quickly because the meter was always running, with movie stars getting paid handsomely to sit in trailers until shooting could begin. "So you learned to hit the ground running and just start hacking away at a problem," he says. "I used to think of it as the Rambo school of design."

In 2002, Brandis, a fifty six-year-old widower at the time, was approached by a friend who'd been doing volunteer work with the Peace Corps in a small village in Mali, Africa. The village water treatment system needed fixing: Would "Rambo" come and lend a hand?

Brandis visited and quickly solved that problem, but then noticed another. "I saw a lot of women in the village shelling sun-dried peanuts by hand," he remembers. The peanut crop was the lifeblood of this village and there was pressure to get as many peeled nuts to market as possible. "These women were doing this all day long, and that is harder than you can imagine," Brandis says. "I could see that their fingers were bleeding." Brandis inquired as to whether it might not be a bad idea for the women to use some type of tool or machine to make the job easier. The response was along the lines of: *Why, that's a great idea—how soon can you get one for us?*

"I have since learned," Brandis says, "that this is how things get started. You notice there's a problem. You think, 'Surely, someone must have invented something that does this little job—I'll go find it.' Then you discover it doesn't exist. And then you start thinking, 'Well, how hard can it be to make one?'"

And so Brandis spent the next year trying to design a peanut sheller. First, he had to "go deep" into the world of peanuts; he sought out the expertise of the only peanut expert he was familiar with, writing a letter to former U.S. president Jimmy Carter. He was referred to the Carter Library in Georgia, where Brandis met a peanut expert who showed him sketches of various inventions designed to shell nuts. None were particularly successful, but one design, from Bulgaria, had a cone shape that sparked an idea.

Brandis designed a prototype that was essentially a cone built within a larger outer cone, with a crank attached so that the inner cone could be rotated. When a peanut was dropped into the contraption and the crank handle was turned, the nut would spiral down between the two cones, which were angled so that the space between them narrowed as the nut rolled down, producing enough friction to cause the shell to gradually disintegrate. Brandis had to get the angles of the cones just right: Peanut shell removal is a delicate matter, because you don't want to remove the nut's brown parchment covering beneath the shell, which serves as a necessary bacterial barrier. Brandis also had to deal with the mysteries of centrifugal force. He'd assumed the cones should be wide at the top and narrow at the bottom, but the nuts rolled better on a narrow-to-wide track—so Brandis, literally, had to turn his original idea upside down.

Brandis's creation was truly a smart recombination, in that he took some existing ideas—involving cone shapes and crank-operated devices—and connected them in a novel way. But where he really jumped the fence was in his choice of materials. Others had tried to carve and shape nut shellers from malleable substances such as metal or wood; Brandis wanted to make his cones from stone. "People thought that was ridiculous," he recalls. But Brandis was doing what any forward-thinking designer should do: enlarging the problem and trying to develop a bigger, more far-reaching solution. He rea-

The double-cone system in version 1 of **the universal nut sheller** looked logical, but the sheller worked better in version 2.

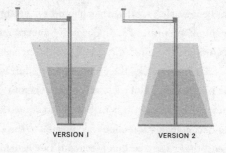

VERSION I   VERSION 2

soned that if he made his sheller from molded concrete, he could then give the villagers not just the sheller but the mold, too—so they could easily make more shellers of their own just by mixing up some cement.

When he finally had a working sheller, he returned to Mali and donated the device, and the mold, to the village. Using the device, nuts could be readied for market fifty times faster—which caused productivity to jump. And no fingers were bloodied along the way.

Brandis estimates there are now some two thousand shellers in use, not just in Mali but also throughout Africa and Asia. He chose not to patent the idea, because he wanted people to be able to use his basic design and molds to make their own shellers, perhaps adapting and improving them along the way.

And that is exactly what happened, as local designers began adding their own wrinkles. "One guy wrapped rubber around the rotor and began using it to husk coffee," Brandis says. Others were using the device to shell pecans, hazelnuts, and Brazil nuts. It turned out Brandis had created a "universal nut sheller," as it is now known. One of the latest twists has come about recently in India, where someone figured out that if you used the sheller to press the oil from the Jatropha plant, you end up with a source of biofuel *and* a good fertilizer.

While continuing to try to get the sheller more widely distributed, Brandis is also designing other low-cost tools and machines for the developing world—primarily using his material of choice, concrete. "I'm basically doing what people did five thousand years ago, when they made things by rubbing stones together," he says. He's created a small nonprofit group, the Full Belly Project, with a couple of former Peace Corps volunteers and a shoestring budget (though Brandis's fortunes did get a boost in the fall of 2008 when he won a $100,000 prize given by the group Civic Ventures to people over sixty who are helping society). But even without travel money, Brandis manages to do virtual globe-trotting on behalf of his creations. "I'm always on the phone or on Skype, telling somebody in Uganda, 'Okay, this is how you use the mold, and then you take this piece and put it together with that piece,'" he says. "I'm doing a lot of long-distance design."

Brandis was able to have an impact on the lives of significant numbers of people in places halfway around the world—even though he had no budget to work with, no materials other than what he could scrape together himself, and no manufacturing or distribution support to help him get his idea made and shared with others. He had to do it all the hard way. Did he ever feel hemmed in by all of these constraints? "No, just the opposite," he says. "Constraints make you more creative, or at least that's how it works for me. If you've got all kinds of options available to you, then how do you know what to do or where to begin? But if all I've got to work with is some wood and cement and maybe a bicycle wheel, I'm ready to go."

BRANDIS'S RESOURCEFUL "Rambo" approach to design is a good model for anyone trying to solve problems in the developing world, where, oftentimes, needs are many and resources scant. But more and more these days, the need for designers to embrace constraints—to take on challenges that call for "doing more with less"—is becoming universal. In the current landscape, where are resources *not* limited?

As economic conditions everywhere have come under pressure over the past year, there's less money to address problems that loom larger than ever. Infrastructures are in need of modernization and social services must be redesigned—but any efforts to do so must work with limited means. In the business world, there's increased pressure to design products and services for consumers who have more complex needs and less money to spend. And at the same time, those products must also be designed with greater sensitivity to environmental impact. Make it cheap, make it sustainable, make it healthy, make it accessible for poor kids around the world, make it fix all of our problems: That's the assignment for design in both the developing and the "developed" world of today.

It's a daunting challenge, to be sure. But if one turns the whole concrete cone upside down and begins to look at constraints the way Brandis and some other designers do, those impediments can be seen as motivators—spurring more practical, resourceful, and innovative design. Referring to the economic and environmental pressures of early 2009, Paola Antonelli, senior curator of design at the Museum of Modern Art in New York, told the *New York Times*, "This might be the time when designers can really do their job." Why? Because, Antonelli said, "What designers do really well is work within constraints, work with what they have."

The term *constraints* is actually used a couple of different ways in design discussion. In one sense, it can refer to the various types of restrictions or barriers that designers themselves build in to the devices or systems they create, in order to prevent people from doing the wrong thing. A warning sign is a design constraint, as is a door that opens only toward you (and thus keeps you from blindsiding people on the other side). But the term is also used more broadly to refer to the limitations, restrictions, and requirements inherent in a particular design challenge. As in: *It must, at least, do this; however, it must not, under any circumstances, do that; it cannot cost more than X; it mustn't take longer than Y to get it done; and it would be really good if it could fly, though no one has ever been able to do that before.*

While constraints do tend to make the designer's job more difficult, they also set the parameters that can guide the creative process and provide a sense of purpose and direction, notes George Kembel, cofounder of Stanford University's d.school design program. Too much freedom, it turns out, actually can be one of the worst constraints for a designer: "What's interesting is that if you give someone an unbounded challenge and say, 'You can design whatever you want, with no restrictions,' it then becomes hard to find a place to get your footing," Kembel explains.

On the other hand, give a designer like Yves Behar or Dean Kamen a set of impossibly severe limits or restrictions—*it must be produced for less money than is humanly possible; it must achieve six functions with one button*—and it only strengthens their resolve. "For me, it is a big part of what makes design exciting," Behar says. "If somebody says to me, 'We've got to design a hundred-dollar laptop, and no one has ever done this before, and in order to get there we're going to have to approach this problem completely differently'—as soon as I hear that, I start believing that we're going to do it."

8.2 "EPHEMERALIZATION": IS IT FINALLY HERE TO STAY?

One of the good things about extreme constraints is that even if you don't overcome them—Behar's group never was able to get all the way down to that hundred-dollar level—you can innovate just by trying and getting close. After the XO laptop broke new ground in 2007 by coming in under two hundred dollars, a wave of other low-cost laptops and netbooks began to appear on the market. So Behar's team succeeded in setting a new standard, but that original goal of a hundred dollars was still out there, beckoning. Interestingly, though, when the One Laptop Per Child group announced plans for the introduction of the second-generation XO-2 laptop, they set a new price goal, seventy-five dollars, which was even lower than the one they had failed to reach.

There are two ways of looking at this. You could say that the OLPC group didn't learn its lesson about being more realistic and tempering expectations. Or you could say that they *did* learn that it never hurts to shoot for the moon and see how near you can get. In any case, by the fall of 2008 Behar's design team was already deep into the second iteration of the little green laptop, and they seemed to be trying to figure out how to do a lot more for significantly less.

Behar had shaken off the criticism from the first go-round. He says that the most important feedback he got was not from the press or critics in the design community, but from teachers and children halfway around the world who were using the laptop. He received images and letters with heartfelt stories, and he stored them on his own computer and continually referred to them for inspiration. Still, he was well aware that the laptop hadn't reached as many kids as hoped. There were a number of reasons why, but the price was surely one of them. So now, Behar and his colleagues were determined to take a second crack at it.

In design, the term "satisficing" refers to settling for what is good enough instead of pushing on toward an optimal solution that might be too difficult or costly to achieve. The OLPC group could have opted, the second time around, to satisfice its way down to that seventy-five dollar mark, just by removing some functionality or using lower-quality parts and materials. Instead, Behar was going the other way—trying to make the computer significantly smaller, lighter, easier to use, more energy efficient, and much more versatile, at half the price.

What magic ingredient might enable Behar to pull off this trick? There were a number of them, actually, but none more important than time. Just in the time that had elapsed since the first design, the possibilities available to Behar had expanded: new materials, new technology, and new ideas. And those options could be expected to keep growing as he continued working on the computer throughout 2009 and into 2010. Behar was counting on "ephemeralization," a design phenomenon described by Buckminster Fuller three-quarters of a century ago.

Fuller believed that ongoing advances in technology, if properly harnessed and utilized, could provide the opportunity for designers to "do more with less"—to achieve more functionality and affordability in designs while using less energy and fewer materials, and generating less waste. By doing that, Fuller believed, designers could improve the lives of people around the world, without draining or diminishing the planet's finite resources.

Ephemeralization once may have seemed like a distant hope for the future, but today it's not so far-fetched. According to Bret Recor, one of Behar's chief designers on the laptop project, "We're at this place now where, while you're designing, you can think about technology or materials that don't quite exist yet, but you know that they're coming soon. New possibilities, at lower costs, are becoming available so fast that you just factor it into your design." This gives Behar and Recor the freedom to ignore present-day realities as they plot the overthrow of constraints. They can design something that is, for the moment, impossibly light, small, and inexpensive—banking on the expectation that by the time the product is finished, technology will have caught up with their dreams.

BEHAR ENVISIONS the new laptop as a kind of magic book that contains the whole world between its covers. When the computer is opened, there are screens on both the top and bottom panels, and as you turn the computer vertically, what you have is, in effect, a left and right page (although when you flatten it into a tablet, those two screens come together to form one continuous surface). "It harks back to the traditional way to read a book," Behar says. "Part of what still makes books beautiful is just having an illustration on one side and text on the other. We wanted to offer that experience. In a way, this kind of re-specs the original format on a book."

One of the important ways this changes the experience of using the laptop is that it allows children to use it in more situations, and also to collabo-

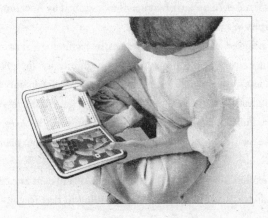

rate more easily. Since the computer is smaller, lighter, and booklike, there's no need to have a table or a desk to rest it on; the designers envision it sometimes being used on the floor, with kids sitting around the "tablet" or passing the "book" from hand to hand.

There is no longer a keyboard on the computer because it now has a touch screen interface. This is a big step forward because it eliminates the need to create separate keyboards in all the different languages of the world. With touch keyboards, OLPC can preprogram the software so that the keyboard updates to whatever language is appropriate. And all of this cuts down on parts and weight—no buttons, no track pad, everything tucked away inside.

Also gone are the cute rabbit ears—evolution (with a boost from ephemeralization) has made them unnecessary, as antenna technology has quickly advanced to the point that it can now be incorporated within the housing of the computer. And as for that housing, it, too, is on the verge of extinction. Behar is in the process of completely reinventing the shell of a computer—which, it was long assumed, had to be some kind of hard plastic because, well—*that's what it's always been made of, right?* But Behar and Recor have posed the stupid question: *Why can't a laptop be more like soft luggage?* They are now prototyping a version that utilizes state-of-the-art vinyl and ballistic nylon to cover the chassis of the computer in a soft shell that is extremely durable. When the computer is closed, it looks like a fabric book, with a carrying ring and tether attached, so you can sling it over your shoulder like a camera bag. But if you should happen to drop it, no problem—as long as you're not an extraordinarily tall child. The precise level of drop forgiveness being designed into the laptop is five feet; Behar figures that is sufficient, so he builds to that spec (and throughout the prototype process, the computer is repeatedly being dropped from that height to see if it holds up).

If it's true that a new design creation is, to once again use John Thackara's useful phrase, a "smart recombination," then that term could also be applied to each prototype that leads up to the final creation. Every new version worked up by the designers is a remix or recombination of elements that were in the previous version, with new elements added. The bits and pieces are moved around and reconfigured—*Okay, this time let's try putting the battery over here, and the fan over there*—and fresh materials, such as the vinyl on the new laptop, are stirred into the blend. Any elements deemed unnecessary fall away to free up room for the new: So long, rabbit ears—you were cute, but we just don't need you anymore.

One of the most important ingredients added to the mix is feedback. The designer must try to incorporate lessons learned from earlier prototypes. Often, they're surprising. Recalling the first-generation laptop, Bret Recor says that what caught the designers completely off guard was the ingenuity of the kids. "The older ones, around twelve, wanted to go inside and explore the components, maybe swap out the motherboard, or fix problems as they came up," Recor says. The designers, meanwhile, had built in a constraint—they'd hidden the screws that kept the shell closed, so the computer would look better and also so it would be harder to tamper with. In doing so, they made it more difficult for the kids and their teachers to service the machine. In the new version, the laptop is now designed to encourage owners to explore its innards, in order to learn about it, fix it, and customize it.

Before the XO-2 was even finished, Behar started working on another iteration of the device that scaled things back further—to a tablet form that was not much more than a screen surrounded by a thin rubber gasket. This newest version, XO-3, is a more demanding design but the target price is still $75 (the plan is to introduce it in 2012). As with the XO-2 prototype, Behar has gone public with all XO-3 sketches and plans, right from the outset. "We're breaking all the rules of industrial design," he confides. "We're revealing what we're going to do two years before we launch it—no company in its right mind would ever do that." So why is he doing it? Why set himself up for critics, who may later point out that the end product doesn't live up to the advance hype? Because Behar believes it is important to make hope visible. He is using what he calls "the tricks of design"—the ability to look ahead, to prototype, and to create desire for something that doesn't even exist yet—because "we want to make everyone believe that this *should* be the next experience," he says. "That is how we'll continue to build a movement for laptops in the developing world. So you could say we're acting like bad boys, in order to try to do something good."

8.3 RULE NUMBER ONE: "MAKE IT CHEAP AS A CHICKEN"

If Behar's prototype provides a glimpse of the future of laptops, his working process—directed toward finding new ways to make something less expensive, more ruthlessly efficient, and more accessible to a greater number of people—may point toward the future of design. In the past, and particularly in the recent past, much of the design industry's efforts have been focused on serving the needs of the few and the affluent. But as a recent traveling design exhibit in the United States argued, it might now be time for a blossoming of Design for the Other 90%. That show featured Behar's laptop and the LifeStraw handheld water filter—both of which were also featured in Mau's Massive Change show—along with an array of other, often startlingly simple creations designed to make life better for people of limited means. There was the Q Drum water transport device, a super-low-cost prosthetic foot, and remarkably cheap and efficient irrigation devices for farmers in the developing world. Jock Brandis's universal nut sheller wasn't there, but it would have fit right in with the show's theme.

This type of design work normally doesn't draw much attention, at least not until lately. But as interest in social responsibility has been growing among designers, the business community simultaneously has begun to be more intrigued by emerging markets, or the so-called bottom of the pyramid. If design is about serving human need, then the developing world offers endless opportunities to fulfill that purpose. And even if design is only about appealing to new customers, well, "a billion customers in the world are waiting for a $2 pair of eyeglasses," noted Dr. Paul Polak, one of the organizers and stars of the Design for the Other 90% show.

It's not an easy market to design for, because the constraints are so severe.

One of the designers featured in the Other 90% show, Martin Fisher, who helped create the KickStart line of water pumps and other low-cost tools aimed at increasing small farmers' productivity in Africa, has seen firsthand the promise and the difficulties of this type of design work. Fisher has had to deal with everything from a complete lack of manufacturing and distribution infrastructure to cultural constraints that can be hard for outsiders to fathom. Once, after a farmer in Kenya began using one of Fisher's pumps and started having immediate success with his crops, some of his neighbors began to suspect "that he was consorting with the devil," Fisher recounts. There were numerous unexpected complications such as this, but above all Fisher had to grapple with what is invariably the number-one constraint when designing for the developing world: affordability. After a number of years observing market conditions in impoverished regions, Fisher came to the following conclusion: A designed product must sell "for not much more than the price of a chicken in the local market."

Fisher's pumps, which sell for as little as thirty-five dollars, are simple in design, relying mostly on the human body to power them, but they can end up revolutionizing a farmer's business (see facing page). More than fifty thousand new microenterprises have been started in East Africa using KickStart equipment, and eight hundred more are started each month. Together, they generate more than $50 million per year in new profits. By one estimate, KickStart pumps have helped more than four hundred thousand people escape from poverty. As Fisher said, "We're not just designing a cool gadget here."

But in fact, Fisher has designed a pretty cool gadget (he had help from IDEO in some of his work) if one gauges the "cool" factor based on the ability of a designed object to do what nothing has done before. The fact is, nothing has monetized small African farms quite like KickStart pumps. Likewise for a number of other simple yet amazingly effective devices being designed for emerging markets. If innovation is defined as a creation that brings about radical change, then the locus of innovation may be shifting to developing areas of the world, where extreme constraints are inspiring designers to be more resourceful and creative. The late C. K. Prahalad,

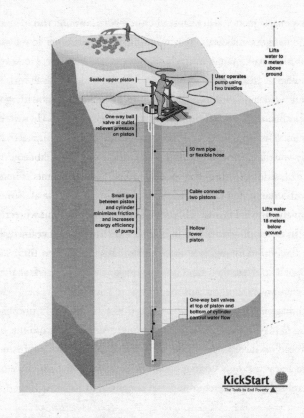

Lifts water to 8 meters above ground

User operates pump using two treadles

Sealed upper piston

One-way ball valve at outlet relieves pressure on piston

50 mm pipe or flexible hose

Cable connects two pistons

Small gap between piston and cylinder minimizes friction and increases energy efficiency of pump

Hollow lower piston

Lifts water from 18 meters below ground

One-way ball valves at top of piston and bottom of cylinder control water flow

KickStart
The Tools to End Poverty

author of *The Fortune at the Bottom of the Pyramid*, observed that designers in emerging markets are often forced by severe price constraints to try to improve efficiency by 90 percent instead of the usual 10 percent—and under such conditions, the designer is required to completely rethink the conventional approach to designing or making something.

Severe constraints can also lead to radical innovation in the use of materials, pushing designers to seek out new ways of utilizing inexpensive ones so as to do less with more or, in the words of architect Shigeru Ban, to create "strength from weakness."

Ban discovered, some two decades ago, that he could build more cost-efficient displays for exhibitions if he made models of buildings from the cardboard tubing that was lying around his office, as opposed to using more expensive wood. Years later, when war-torn Rwanda was in crisis and needed emergency shelters, Ban was watching the images of refugees on his television set in Japan and felt moved to do something. He knew that public officials and humanitarian agencies trying to deal with the crisis were strapped for resources and couldn't afford to build shelters with wood or metal. So Ban stepped forward with a surprising proposal: He would offer to help build housing in the area using cardboard tubing as his material. The agencies were uncertain at first, but when they learned that Ban could build housing units for as little as fifty dollars apiece, "they welcomed my idea for a more affordable alternative," he says.

After the Rwandan shelters proved sturdy, Ban began to erect paper houses in other disaster areas, including once in his own country, Japan, during the period following a major earthquake in Kobe in 1995. Initially, for people who'd lost their homes to the quake, he constructed what looked like elegant log cabins—except the logs were paper tubes. Then, when he learned that one of the damaged areas had lost its church, "I proposed rebuilding *that* with paper," Ban says, adding: "At first, they did not believe that I could do it." Oh, ye of little faith: Ban's cardboard church stood for the next ten years, only coming down recently when the paper walls were disassembled so that the church could be moved to Taiwan.

8.4 RULE NUMBER TWO: THE CHICKEN SHALL LIVE FOREVER

While cost constraints compel designers to seek out new ways to "make it cheap," growing environmental concerns are placing perhaps an even tougher demand on design: "make it last forever." This is an area Mau has begun to stake out in his work with major companies on sustainable design. If you think of it in a narrow sense, sustainable design places limits or constraints on designers because it dictates that you must not (or at least should not) use certain materials that are harmful to the environment and, further, that you must not engage in processes or practices that may contribute to waste or contamination. This discourages or removes certain design options; ergo, constraint.

Mau, on the other hand, tends to see sustainable design as an expansion of the old design process—a more holistic way of thinking about it. "For most of design history, we took organic material and we shaped it, and that shape was the design," Mau says. But that shape is really just a temporary form, Mau points out; the matter used to make a glass had its own design before it was shaped into the glass and will take on a new form and a new destiny after the glass breaks. What designers must do now, Mau says, is begin to ask more questions about where that material is coming from and what will become of it after it has ceased to serve its function.

"Right now, there is a vector that goes from raw material to designed object to the garbage dump," Mau says. The way to change that is for designers to focus not just "on the temporary object, but on the continuous flow of matter. We have to design that flow. We have to change that vector, turn it into a continuous circle." (Mau has boiled this concept down to a slogan, which he shared at a recent business conference: "We're no longer designing for the present tense," he said. "The new design is forever.")

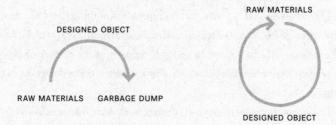

Mau is echoing the call that has been issued by the sustainable design guru William McDonough, author of the renowned book *Cradle to Cradle*, which explained that designers have been making objects destined for the "grave"—meaning a landfill or incinerator—when they could be designing things that return to the cradle, reborn in new forms. To do this, says

McDonough, all design materials must be looked at "as biological nutrients that can safely go back into the soil, or as technical nutrients that can go back into technical cycles and become raw material for new products." It requires a complete rethinking of the design process, he says: "To eliminate the concept of waste means to design things—products, packaging, systems—from the very beginning in the understanding that waste does not exist." Which means that when making any product, McDonough says, the designer must always ask: "Does it have reverse logistics? Do you have a way to get it back to soil or back to industry?"

One example of how this can work can be seen in the efforts of the carpet maker Shaw Industries to create cradle-to-cradle carpeting. The company has used carpet materials designed by McDonough that can be reused as carpet fiber on its new products, and has also worked with Mau on designing some of its products. But perhaps most important, Shaw has designed an operational system to make possible the "continuous flow" that Mau and McDonough are talking about: It involves stamping an 800 call number on the bottom of the carpeting so that when people are done with it, it's easy for them to call the company and get it taken away to be recycled by Shaw. If the whole process is designed well and efficiently, the company can actually save money in the end—as Shaw has.

The cradle-to-cradle phenomenon represents nothing short of a revolution for design because, as McDonough states, "It means that *everything* has to be redesigned. So it's the largest job creation in the history of design, larger even than modernism, which also called for the redesign of everything."

The revolution even has its own charter, with more than a hundred thousand designers signed up. It's called the Designers Accord and was authored by IDEO designer Valerie Casey as a kind of shared promise among designers that they will apply a much more eco-conscious mind-set to everything they design—and that they'll also urge their corporate clients to do likewise. Casey says that the accord represents an opportunity for designers to atone for past mistakes: "We've been creating bad stuff for a very long time," she says.

BRUCE MAU

"Yes Is More"

When it comes to changing behavior, we have fifty years of evidence that going negative doesn't work. For over half a century, environmentalists have scolded us to "reduce," "use less," "give up" this or that, and, most pointedly, to "get out of your car!" Over all those years, the total number of cars in the world inexorably increased. Last year alone we produced roughly sixty-six million new cars—adding four times as many cars to our roads as we did in the sixties.

Instead of rejecting it, we embraced the car and its intoxicating effects as never before. Around the world, many cultures and countries may not have fully embraced human rights, freedom, and secular democracy—but they have embraced traffic. The few remaining outposts that have yet to get cars in large quantity are desperate to have them.

During the last fifty years, we used most of our innovation and advancement in energy efficiency—about 1 or 2 percent per year—not to make cars lighter and cleaner but to make them bigger and more powerful. En masse we went in the wrong direction. As the bicycle-riding environmentalists scolded, we closed our power windows and turned up the air-conditioning.

When it comes to changing attitudes about the environment, apparently "No!" is not the answer we were looking for. Getting hit with a green stick has had little effect. So how do we convince the six billion-plus people on the planet that changing the way they live is critically important to their future?

Think orange. Think carrot, not stick. Seduction, not sacrifice. Yes!, not No! If we are to accomplish the objectives of the environmental movement—to create a culture that can exist in perpetuity and in harmony with the ecological systems that support us—we must reimagine and redesign everything we do. But we must also do so in a way that allows people to experience beauty, exhilaration, love, pleasure, and delight without destroying our planet and its nature.

There is only one way to make this happen: Use the power of design to make the things we love more intelligent. Embrace the revolution of possibility that we are living through, to radically reduce the material and energy we use, while increasing the positive impact and effect of the things we use in our daily lives. We will make the new sustainable ways more compelling, more attractive, more exciting, and more delightful than the old, destructive, short-term ways. We will compete with beauty, and make the smart things sexy.

So far, we have failed in designing a real alternative to the car. When you compare the bus and the car as experience, there is a clear winner and loser. Why does my minivan have seventeen cup holders—but my bus has none? Why is my bus shelter not heated, but I can start my car remotely and let it warm up? Why is my bus uncomfortable and noisy when I can listen to Beethoven in my car in relative silence? My bus is a design failure. It's a stick painted green, and out of desperation or inspiration, I'm supposed to want the experience. In Toronto, the slogan of the transit company is "the better way." Well, actually, no. It's not the better way, and everyone knows it.

Until we design a bus experience that is more attractive, more effective, and more elegant than the car, we will be selling a losing proposition. The same applies to the car itself. We must imagine and redesign the car as a product with positive impact and not make our design objective a car that is less negative. We must design an ecology of movement options that are thrilling in every way, and that also fit together as an ecological, sustainable—but most important, sexy—system.

If we are ever to achieve the ambition of the environmental movement, we have to get beyond "No!," face the problem directly, and define what "Yes!" would look like—and not simply continue to hope that one day we will somehow collectively wake to a world of altruistic people who reject the car. "No" is not the answer. Yes is more.

WHILE THE challenge of "designing for forever" is relatively new to most designers, both Mau and McDonough like to point out that nature has been employing this method of design for, well, forever. It's one reason why the sustainable design movement has been crossing paths with the field of biomimicry, which studies natural design and tries to extract lessons and principles that can be applied to man-made design. Mau is a great admirer of Janine Benyus, who literally wrote the book on biomimicry (*Biomimicry: Innovation Inspired by Nature*) and who was featured in Massive Change. Benyus has suggested that as designers try to grapple with various constraints—from making products and systems more sustainable to just making them more efficient and productive—a lot can be learned from studying the way nature designs solutions.

In fact, thinking about nature can be a starting point in design, Benyus says. When designers are asking those initial "stupid questions" such as *What should my product do?* and *How do I make that happen?*, Benyus advises that they should also ask, *What in the natural world is already doing what I'm trying to do?*

"Chances are that you'll find not one natural model but several," Benyus says. Biomimicry is particularly good for mimicking form—"for looking at the incredibly elegant and efficient shapes in nature and what they are able to do"—as well as processes (like photosynthesis) and entire ecosystem survival strategies.

Benyus has worked with designers and engineers at companies such as Boeing, General Electric, Herman Miller, and Procter & Gamble, who typically come to her with a particular design challenge that has its own set of constraints. Benyus then tries to point them toward relevant design scenarios in nature. For example, she worked with Nike as the company was developing apparel for the Summer Olympics; Benyus showed the company's designers how the African reed frog manages to stay cool and comfortable in its own skin, even in intense heat conditions. "In every habitat, there's a set of opportunities and a set of limitations and those are the parameters that organisms dance within," Benyus says. "These organisms have developed certain design solutions that take full advantage of those opportunities and operate with full respect for those limits. And to me, that's very similar to what a designer does."

It can be difficult to know where to look for relevant information in nature or how to sift through all the instructive possibilities, and to that end Benyus has created a kind of "Google of nature" in the form of a Web 2.0 portal called AskNature.org (a collaboration with a larger project, the Encyclopedia of Life). Gathering together information from biologists around the world, Benyus organized it all by function, so that when a designer goes to the search bar and types in, "How would nature . . . sense vibration?" or "How would nature . . . store gases?," real-world examples pop up.

With regard to sustainability, biomimicry is particularly chock-full of good lessons, because only sustainable design models survive in nature. So, for example, as we endeavor to design solutions to the challenge of water shortages, we can learn from organisms like the Namibian beetle, Benyus notes. It has bumps on its wing scales, which lift up into windblown fog, and those bumps have water-loving tips that are designed to pull water from the air and collect it. By mimicking the way those wings work, we could turn fog into potable water.

Natural ecosystems also hold many valuable lessons on upcycling: What would appear to be waste from one organism becomes food for the overall ecosystem. Which makes Benyus wonder, "What if you could send back your iPod to Apple, which would then take those same materials and re-manufacture them into new products?" If you signed up to be part of the "Apple product ecosystem," you might own a limited pool of materials that would be cycled over and over again within the Apple product ecosystem, she says. "Therefore, I'm not just getting a cool computer, but I'm also contributing to a whole new kind of economy that makes the most of its materials." Apple could take your old stuff and recombine it with new parts, or just "put it in an enzyme bath, break down those materials, and use them to make some other product," she says. "This is what ecosystems do."

8.5 HOW TO MAKE A CAR THAT BREAKS DOWN

A cradle-to-cradle iPod is not a bad idea (are you listening, Apple?). But a cradle-to-cradle car—*there's* a way to make a dent in the world's scrap heap. In the fall of 2008, the car designer Gordon Murray, who works independently of any major car company, operating from a modest sixteen-person studio just outside of London, was putting the finishing touches on a prototype for a car that would be mostly recyclable, incredibly small and fuel efficient, and extremely affordable. He seemed to be going for a trifecta in terms of defeating car design constraints.

The odd thing is, given Murray's background, he is not someone who might be expected to take on such a constraining assignment. Murray is widely regarded in automotive design circles as a genius when it comes to designing cars with *no* limits: ones that go incredibly fast and cost considerable sums of money. His multi-million-dollar McLaren F1 supercar is legendary, though it's just one of a line of Grand Prix–winning race cars that Murray has designed over the years. But when it came time to design his twenty-fifth car, which he is calling the T.25, Murray shifted gears.

He recalls that it all started when he was sitting in traffic on the A3 just outside London one day. "I was looking around me at all these stationary vehicles with one person in them—big BMWs and 4x4s—and I just thought, 'This is really not sustainable. This can't go on.'" So Murray set about designing an "urban personal transport" car that *would* be sustainable—one that addressed issues of traffic, parking, fuel economy, environmental impact, and manufacturing efficiency. He started sketching ideas right away. He knew from the outset that the car had to have a tiny footprint in order to address congestion and efficiency issues, so he set out to make one about half the size of a Mini.

Murray wanted to approach the design with "a completely clean sheet of paper"—not only in terms of this particular car, but the whole process of making cars. One of the big "stupid questions" he raised: *Why doesn't a car come apart?* What if you assembled one such that it could be easily

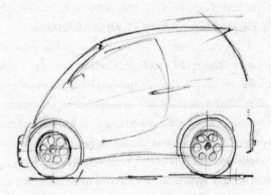

*dis*assembled by an owner, armed with, as Murray says, "nothing more than a spanner" (also known as a wrench)? Parts of the car could then be returned for new parts, with old bits being reused in the manufacturing process.

Murray has designed the T.25 so that there are no welded, fixed, or bonded elements. His design used low-cost composite materials throughout the car—the removable panels that comprise the T.25's body are made from upcycled plastic bottles. If one of the panels becomes dented or damaged, no problem—take it off and replace it with a new panel. ("Having in the past been presented with large repair bills for tiny dents, this was important to me," Murray says.) And if you want to replace it all, "you don't chuck the car—you rebody it," he says.

The biggest constraint in making the car is cost. "It's all very well selecting exotic lightweight materials for the McLaren F1 because you can afford it," he says, "but if you're trying to build a car for five and a half thousand quid retail [less than ten thousand dollars] it's a huge constraint." The second biggest constraint is the size, "because you may be making your car smaller," he says, "but you must deal with the fact that *people* aren't getting any smaller."

This is the part of design, Murray confides, that he really loves: maximizing the internal "packaging" of a car, getting the most out of the available space. Just like Yves Behar, he does it by moving around components, eliminating anything that's not needed, splitting the radiator into two pieces if it helps (as he did in the design of the McLaren). On the T.25, the greatest space saving was achieved by moving the engine to the rear of the car and canting it at over 45 degrees. By the time he was finished moving things, he'd somehow freed up enough interior space so that the car, though much smaller than a Smart car, can seat twice as many people (four).

Even though Murray was designing a sustainable car, he opted not to make an electric one. He says he felt it was important to design a car that people would actually want to drive. "There's some awful little electric cars buzzing around London and you wouldn't want to be seen in them," he says. "When one of them goes past, the feeling is 'What the hell was *that*?' You've got to have the opposite—like the original Mini, something people want to be seen in."

Murray intends to license the rights to the T.25 to a manufacturer who can then follow his detailed specs, right down to the design of the factory where the cars are made (Murray has worked out every detail in the making and servicing of the car, with an eye toward minimizing carbon emissions each step of the way). If all goes well, the T.25 may be on the roads soon, though who will make it and what it will be called remains to be seen. But you'll know it when you see it—just look for a car that fits two to a lane or three to a standard parking space.

8.6 ALL TOGETHER, NOW

When Gordon Murray talks about constraints, he sounds a lot like Yves Behar or Jock Brandis: He speaks about the thrill of doing what hasn't been done and the challenge of rearranging the pieces of a design puzzle until an "impossible" problem is solved. Murray says he is forever searching for and testing new materials so that he can lower a car's weight without raising its price. And he spends a good deal of time jumping fences—taking design ideas from outside the realm of cars. He's found car design solutions by studying, for example, the design of helicopters. "As a designer, you must have a roving eye," he says. "You have to think laterally, and cross-feed ideas from everywhere."

The good news for designers is that there is more to "cross-feed" upon than ever before: more information and ideas, a mouse click away; cheaper, more efficient materials and smaller, more capable components being developed constantly. There has never been a better time to be a smart recombinator. Bucky Fuller himself, who died before the Internet revolution, could hardly have imagined the current possibilities for doing more with less.

The challenge for designers, however, is to harness and sort through those infinite possibilities. Ideas, knowledge, or raw materials that can help solve just about any problem or deal with any constraint can probably be found out there somewhere—and if you're lucky, as Brandis was, you might happen to stumble across an obscure idea from Bulgaria. But the likelihood of connecting with the right ideas and resources increases greatly when collaboration comes into play.

In Massive Change, Mau described the new collaborative environment for problem-solving, made possible in part by connective technology. Design is no longer about "one designer in one place trying to come up with one solution." Instead, Mau declared, "Problems are taken up everywhere, solutions are developed and tested and contributed to the global commons, and those ideas are tested against other solutions."

In many cases, the constraints designers are dealing with are universal problems: *How do we make it easier to do this one very important activity?* Or *How do we make this critical resource more available and accessible to people who really need it?* When the challenges are this fundamental and important, chances are someone, somewhere has been thinking about them and maybe even doing something about them.

When Cameron Sinclair of Architecture for Humanity launched his Open Architecture Network (OAN) in 2006, creating a platform for designers around the world to share ideas on how to deal with difficult conditions and severe constraints while building housing for people in need, he was stunned by the response. Practically overnight, he had more than a thousand design ideas and blueprints submitted to the network.

And it had a tremendous impact on the work of designers in the field, trying to deal with problems and crises. "Designers who felt like they were alone out there in the wilderness suddenly found they were connected to people who could help them," Sinclair says. "They realized, 'Hang on a minute—there are five other projects out there *also* happening on Native American lands, involving low-cost housing, and using indigenous materials. So let me find out what *these* people did.' It becomes a peer-to-peer networking situation where people are not just sharing their ideas, but their expertise."

Collaborative design efforts like OAN are sprouting up all over. Valerie Casey's Designers Accord is intended, in part, to encourage designers to share experiences and best practices about trying to deal with environmental constraints: *Who's tried what? How well has it worked? Any idea how to take it to the next level?* Since design is an iterative process in which everything must be tested and retested, the possibility of connecting with other experimenters is not only about sharing ideas but also about saving time on unnecessary trials.

It also brings more diverse thinking to bear on a particular problem and set of constraints. Recent studies—such as those by Scott Page of the University of Michigan—have shown a direct link between diversity and innovation: the more people that are focused on a problem, and the more varied the experiences and backgrounds of those people, the better the chances of finding an original solution to a difficult problem.

And once a solution is found, collaboration becomes even more important, because a diverse skill set is usually needed to turn the idea into a reality. Hilary Cottam, of London's Participle design group, has argued that designers today may need to be facilitators above all else: marshaling and integrating the efforts of engineers, sociologists, politicians, social service administrators, community activists—all the people who may play a critical role in bringing about change through better design. The idea is only the beginning.

In the new collaborative environment, the question sometimes arises: *Whose idea is it? Who is the designer?* A few years back, when Cottam received honors as the top designer in the UK for her work on highly collaborative team efforts to redesign a local school system, some critics asked: *But what did she design?* They were judging by the old "solo artist" design standard and clinging to the notion that designers make things, objects. Cottam was dealing with the new, more complex realities of designing systems in today's world. To succeed in this environment, it becomes necessary for the designer to share ideas, defer to the expertise of partners, and check one's ego at the door.

But that's a small price to pay for admission into a new world where the possibilities are multiplied and where ideas have a better chance of becoming solutions. That's the bottom line, after all. As Dean Kamen says, "We have to do whatever it takes to get ideas out there into the world. Otherwise, you're just doing science fair projects."

PERSONAL

9. DESIGN FOR EMERGENCE

Applying the principles of transformation
design to everyday life

LIFE AS A DESIGN PROJECT

In the fall of 2008, Mau left Chicago and spent a week in Florida to start
work on a new design project. He was taking on what may turn out to be his
biggest transformation endeavor to date: an attempt to redesign himself.

The whole thing started when Mau was called to the office of Miles
Nadal, who runs the conglomerate holding company that owns a stake in
Bruce Mau Design. With the economy tanking, Mau was bracing for some
possible bad news or at least some dire warnings about the need to produce
more profit. He was ushered into Nadal's empty office. Then, when Nadal
arrived, Mau recalls, "only two-thirds of him walked into the room."

Nadal had slimmed down after undertaking a complete overhaul of diet
and lifestyle under the supervision of a holistic health specialist. After explaining all of this to Mau, Nadal warned him that because of excess weight,
"you could become diabetic, you could go blind, you could lose your limbs."
Nadal pointed out that his corporation, MDC, had an investment in Mau's
future—"the biggest risk for us is that you drop dead," he said—and he
offered to send Mau to see the same health specialists who'd helped him.
"He basically said, 'I'm going to change your life,' " Mau says. "It was one of
the most amazing gifts I've ever received."

Thus Mau journeyed to Palm Beach and spent several days being prod-
ded, examined, analyzed, and interrogated. One of the things that came out

was that due to extreme apnea, he wasn't sleeping; he woke up more than fifty times an hour. This contributed to his body's craving for energy.

Mau needed a comprehensive plan and a rigorous system for living that involved getting more and better sleep, eating the right foods, exercising regularly. None of this was simple, however, because Mau's lifestyle, with its intense deadlines and constant travel, tended to conflict with these regimented strategies. Mau was dealing with a wicked problem, and that problem was his life.

He approached the problem the way he does most design challenges. He asked the stupid questions, the ones so obvious they never seem to get raised—such as, *What is really important to me in my life?* "I realized I need to simplify and clarify my life," he says. First, he sketched out his life on paper, complete with lists and descriptions of projects. When it was all laid out in front of him, the designer in Mau could see immediately that there was too much there—too many attempts to try to do too many things; too many distractions; everything squeezed together. Mau's life had developed a case of featuritis.

"Once I had everything laid out in front of me, I thought, *Some of these things I've got to stop doing*," Mau says. "Maybe it would be wonderful to sit on that board with those interesting people, but how does it fit in with the larger plan? I have to start thinking that way." On the other hand, Mau does have an insatiable appetite for opportunity. "There are always these wonderful possibilities that come along," he confides, "and how do you decide which ones to say no to?"

A good example was a project that involved redesigning a major Middle Eastern city. It had just come on the table as Mau was embarking on his life redesign. It promised to be difficult: Mau was being asked to help provide a blueprint for the city's future, addressing such issues as traffic congestion, cultural conflict, and the need to balance modernization with tradition. To do the project Mau assumed he'd have to shuttle back and forth on fourteen-hour flights. And just to win the job, he would have to assemble an elaborate exhibition—a kind of mini Massive Change—showing all the possibilities for reinventing the city. This whole show would be done for the benefit of one powerful person (whose name Mau was not at liberty to disclose), who would walk through the presentation, consult with colleagues, and then, presumably, give the thumbs-up or thumbs-down.

Mau debated whether he could handle such a difficult and time-consuming assignment. But then again: A chance to redesign the future of a major city? How do you say no to that? Mau wants to live more healthily, but "I also want to have an exciting life," he says. And this is exactly the kind of thing that makes the problem complex.

One of the things Mau figured out early in the redesign process was that his travel schedule needed fixing. He was traveling round-trip from Chicago to Toronto twice a month, which seemed like a reasonable way to oversee two offices. But when he did the math, he realized he was spending twenty hours a month—240 hours a year, or six full working weeks—going to and from the airport and sitting on planes as he shuttled between offices. The designer in him saw a problem that needed solving, and he began devising a plan to travel less but spend more time during each visit. He also increased the use of Skype to get in the habit of conducting more virtual meetings and collaborative sessions between Chicago and Toronto. They were relatively simple fixes: Mau had just never thought it through before, never asked if there was a more efficient way.

This was just one small element in what would have to be a larger system redesign of Mau's overall life, both business and personal (the two being, of course, intertwined). Mau needed a working plan that would improve health and increase productivity. But he knew that if this plan was to work long-term it had to be appealing, not punitive. It would have to be designed in a way that would make him want to welcome and embrace the changes. It would also have to allow for variation, occasional failure, and unintended consequences. As he set out to begin, Mau realized that he was nervous but also very excited about the prospect of applying massive change to himself.

IT MAY seem strange to think of life as a design project, but when you consider Mau's basic definition of design—the capacity to plan and produce desired outcomes—it certainly is applicable to life planning. The same is true of some of the other design definitions mentioned earlier in the book, such as Victor Papanek's: the conscious and intuitive effort to impose meaningful order. (Who among us doesn't have "impose meaningful order" written on his/her to-do list?) And then there is one of the favorite definitions used by Milton Glaser (though it originated with the sociologist Herbert Simon), which suggests that design is all about *Moving from an existing condition to a preferred one.* That sounds like it could be the subtitle of every self-help book ever published.

One of the dangers of using these broadened definitions of design is that, as Glaser points out, "you quickly get to the point where it starts to seem like every act is design, and you can get lost in that definition because it's too cosmic." So, yes, technically, when you decided what to wear to work this morning, you made a design decision, and you made another one when you decided which route to take to avoid traffic. But while we may constantly make small, immediate life design choices, it doesn't mean we're *designing*, in the larger, more holistic, transformation-inducing sense of the word.

One important difference is that those small, everyday design decisions and actions often tend to be reactive, rather than proactive: addressing individual challenges as they rise to the surface instead of anticipating them and dealing with them as part of a cohesive systems-design approach. Designers are as guilty of this short-term life design thinking as the rest of us, Mau acknowledges. "When I stop and think about it, I have allowed so many things in my life to just happen. Some of those things certainly could have been designed better. And I'm a designer! If anybody should be thinking about these things, it's me."

Mau believes that people are hesitant to even think in terms of designing their lives because it can be intimidating. "I think we're afraid of taking on that responsibility," he says. "What if the design we choose is wrong?" And if one of the first steps of design is to ask stupid questions, well, that can be a nonstarter for some who may not feel comfortable or see the value in asking such questions about their lives.

There's also the issue of *time*, which may be the ultimate constraint for many people when it comes to trying to create a design for life. How do you stop, step back, and design something that just keeps moving forward and pulling you along with it?

So people end up not designing their lives for the same reasons they may be reluctant to design their businesses or to try to design social sector solutions—not sure it's necessary; not sure it'll work; and who has time for design?

But there are at least two compelling reasons why we should consider applying design principles in our lives today: number one, because we can and, number two, because we must. The first gets back to the current phenomenon Mau refers to as "distributed possibility," meaning that we have more creative and analytical tools, more information, more understanding of problems and possible solutions than ever before. With all of those tools and all that knowledge at our disposal, we have more capability to design the way we live, should we choose to try.

As for why we "must," it's based on the sense—and the growing mountain of evidence—that we no longer have the luxury of living without a plan and an end vision and without an understanding of constraints and consequences.

9.2 "A DESIGN FOR LIFE WAS SOLD TO US. AND IT WAS FLAWED."

Brian Collins is adamant on the subject of designing one's daily life. "It's not a question of *should we* design the way we live," he says. "We *have to* start to think about our lives that way. It's imperative." Collins points out that many of the problems we face today, particularly in terms of dwindling resources, have reached a crisis point precisely because, as Collins puts it, "our lives up to this point have been designed unconsciously—it's as if we've created those lives with our eyes and ears closed."

To some extent, the designing was done for us, and not necessarily with our best interests in mind. "The way we live, and how and where we live—a lot of those decisions were influenced by car companies, energy companies, and other industries, including advertising," Collins says (and with regard to that last group, Collins readily admits that he was one of those influencers). These various interests, Collins believes, projected "a vision of the future where gas would always be fifty cents a gallon and there would always be plentiful resources." That was the "hope" that was made visible to us all. "It was a design for life that was sold to us," Collins says. "And it was flawed."

Well, fair enough: It's hard to dispute that some large design mistakes were made. Still, where does all this leave us today? The problems Collins is talking about—global warming, water issues, population growth issues— are huge and seemingly beyond the influence of the individual humbly trying to design a good life.

Don't try telling that to a designer, however. Designers believe that we have the power and responsibility to create the world that exists around us. And they also know, from design experience, that all the small individual decisions made on a journey have consequences. The tiny compromises and moments of carelessness accumulate, and then one day the whole bridge collapses. This may be why designers tend not to respond well to the "what can I do—I'm just a cog in the wheel?" argument. As designers, they happen to know that every cog is structurally significant.

Suggest to Marianne Cusato, the Katrina Cottage designer, that the individual doesn't have much control over the fact that houses have been built too far apart from one another (causing us to drive and pollute too much, while also eroding our sense of community), and she practically screams. "It's definitely *not* out of the individual's hands," she says. "We let ourselves think so and then allow ourselves to live in ways we don't really want to live." Cusato believes the first thing people must do is make a decision: *Do I want to live in a way that is sensible and responsible?* If the answer to that stupid question is yes, they begin the process of redesigning how they live, including where they live. (And if they're really ambitious, they progress to trying to influence town planning groups and others who make design decisions about how we *all* live.)

But you begin with the area in which you exert the most design control. It's in our homes, on our property, and in our daily activities that "we have the freedom," writes the designer Fritz Haeg, "to create, in some small measure, the world in which we want to live."

MAU SHARES that philosophy, and in fact lives it: He doesn't own a car, and when he moved into his house in Chicago a couple of years ago, the first thing he did was to excavate a forty-foot hole in the backyard to install a geothermal system that heats and cools his house without using gas, oil, or electricity. But at the same time, Mau sees the "social responsibility" piece as just one part of the puzzle of designing your life. He also thinks you can design your life so that it is more interesting and enriching. As an "experience designer," Mau believes that life—not unlike, say, his Massive Change show—should be designed as an experience that gathers momentum and becomes increasingly compelling as you "wayfind" through it.

Design is meant to be a tool for change, but Mau thinks the most complex designs can go beyond spurring change in the present—they can also anticipate and build in accommodation for future change. In his dealings with the business world, he sometimes refers to a principle he calls "designing for constant change."

The principle of *Designing for emergence* is similar, but perhaps even more appropriate as a guideline for life design. A number of designers, social scientists, and writers have dealt with the concept of emergence, including the design researcher Greg Van Alstyne, an early and integral figure at Mau's Toronto design studio for a number of years and one of the people who headed up the Massive Change project. Van Alstyne has since gone on to cofound the Strategic Innovation Lab, an advanced design research center based at the Ontario College of Art & Design. He has made a close

Emergence in nature, in the form of a Fibonacci spiral

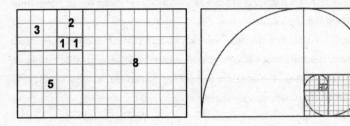

study of design and emergence—the latter having to do with the way organisms and communities grow and evolve over time, often in unexpected and complex ways.

Van Alstyne notes that life itself—that includes each person's individual life—is emergent in the sense that it unfolds in ways that can be surprising and sometimes quite wonderful. The nature of emergence is such that it cannot be fully controlled or designed. But you can encourage it; hence, you can design *for* emergence by providing the right conditions for growth, blossoming, enrichment, and evolution. Van Alstyne notes that if you study the way certain organisms in nature emerge, such as sunflower heads, they follow a spiral growth pattern that has the same structural properties at any scale. This pattern corresponds to a well-known mathematical formula known as the Fibonacci sequence, in which each number is the sum of the preceding two numbers.

Because of this particular growth pattern, sunflowers "can grow to insanely large sizes as they make use of the Fibonacci spiral from the beginning of the growth pattern to the end," Van Alstyne explains. "In fact, you don't necessarily know where the growth will end—it just continues to emerge." Van Alstyne likes to think human life has the kind of emergent qualities seen in the sunflower head, the same possibilities for unexpected and surprising development that continues beyond expectations. "It provides a kind of hopeful moment for these otherwise perilous times," he says.

To design for emergence requires a certain kind of designer, Van Alstyne stresses—one who plans for possibilities but also allows for surprises. In the business world, Google, Facebook, and Apple are examples of companies that are designing for emergence. They create products and services that are released out into the world and then evolve and grow as people engage with them. These companies have shown themselves to be highly adaptable to changing conditions around them. But they also are good at arranging and designing the conditions that allow for their own growth. Apple knows how to create a supportive ecosystem (iTunes, the Apple Store) where its own products can flourish. Facebook is designed so that the community of people using it self-replicates and expands of its own accord. Google is designed so that it never stops learning—it is always learning from the people who interact with it while also continually branching out into new areas of expertise by devouring additional knowledge bases. And all three companies seem to be constantly creating new and slightly evolved versions of their products and themselves. (Remember when Apple was just a computer company? And when Google was just a search engine?)

What are the key lessons of designing for emergence that can be picked up from watching these companies? Here are four shared principles that jump out:

1) Design your immediate surroundings (your ecosystem) in a manner that is self-sustaining and conducive to growth.
2) Develop a strong, supportive relationship with the community around you.
3) Keep learning.
4) Keep creating and reinventing.

As it happens, these are four themes that also seem to come up repeatedly with Mau, Van Alstyne, and many other designers on the general subject of designing a better life.

9.3 CREATING A RICHER, MORE SELF-SUSTAINING ECOSYSTEM (OR, WHY MINIMALISM IS SO OVER)

If we start with immediate surroundings—one's own personal ecosystem—what comes to mind first is the home. And this quickly brings up the subject of "home design," which, of course, is the stuff of endless cable television programs and glossy magazine layouts, all revolving around the remodeled kitchen and the updated master bath. The "design porn" culture, which has made a fetish of countertop materials and must-have household objects, is seductive fun that may or may not be harmless. (Did all those images help fuel the house lust that drove some people into the subprime mortgage quagmire? Who knows, but as Van Alstyne observes, the TV design shows do seem to suggest that "if only you had a more spacious family room, all your problems would be solved.")

Mau's main objection to the home-decorating media is its focus on images of pristine perfection. That, and the fact that it really has very little to do with *your* home; it is putting forth someone else's vision of perfection as a model to aspire to. Mau once wrote a screed about this, which included the following lines:

I'm sick of modern design.
I'm fed up with corporate cool.
I can't bear to see one more "continuous surface."
I've had it with perfection.
I hate clean lines.

"Great design is not reductionist," he says. "A lot of current design seems to be about just getting everything down to a clean, perfect experience—which, at its worst, is stale, cold, and lifeless. I think good design should be complex and rich and even messy, like life. When you think about it, design is a philosophical statement. It's a way for you to say to the world:

'This is who I am. This is what's important in my life, what I value, what I believe.'"

All of which suggests that the inspiration for designing your personal surroundings should come from *within*, not from outside style experts or from the design media. Obviously, decorators and home design experts will take issue with this. It ignites the whole democracy-of-design argument, with one side in favor of "doing your own thing" and the other side extolling the merits of relying on refined tastes and standards. Andrew Keen, a social critic and author who falls into the second camp, says, "Let's face it, most people don't have the talent, the discipline, or the commitment to do their own designing, which is why we hire professionals." (Keen has a devilish sense of humor about all this: "I'm not a fascist. I'm a semifascist," he says. "If people want to design something ugly and nonfunctional, fine, though I'd rather not have to look at it. So let them design their own clothes—as long as they don't go outside.")

But whether you design it yourself or with help, a home does seem to be the place to make a personal statement of values, beliefs, and life experiences. When talking to experience designers on this subject, the word "meaning" keeps coming up. A designed environment—whether it's a museum designed by Mau or an entertainment venue designed by David Rockwell—tends to become more meaningful to people when they get the sense that there is a rich and authentic story unfolding before them, as expressed in many details that come together to form a narrative.

Mau thinks a home can and should tell a story, too. And that story is there for the benefit not just of outsiders but for the people living in the home. When the psychologist Mihály Csíkszentmihályi studied what was most important to people in their homes, in terms of what gave them the most pleasure and even inspired them to be creative, he found it wasn't the latest shiny new design objects. Usually, it was the old trinkets, worn furniture, favorite books—things that, as Csíkszentmihályi noted, evoke memories and personal stories.

A home that tells rich stories about who you are and where you've been can end up being pretty eclectic—and, as Mau notes, it's likely to be more "maximalist" than minimalist, reflecting the complexity of life. From an *Architectural Digest* standpoint, it might be seen as too personal, a bit cluttered, too "lived in." But if you think of a home as part of an ecosystem designed to enable you to live and function comfortably while also providing meaning and inspiration, there's no such thing as too personal or too lived in.

In this context, having "good design" in the home doesn't require spending lavishly on the latest object of the moment as determined by cultural tastemakers (that approach requires constantly discarding the old and buying the new in order to keep updating a design story that was never really yours to begin with). If only to avoid adding to the trash heap, it may be better, as Mau suggests, to create your own design stories based on things that are deeply and personally meaningful to you—and are likely to stay that way.

ONE OF the ways to design a more meaningful personal environment is to instill the home itself with a sense of purpose. It's interesting to see that homes designed to be environmentally sustainable end up providing unexpected psychic sustenance to the people who live there—the home becomes personally meaningful and rewarding, based partly on the design effort and ingenuity that goes into creating and maintaining it. Unfortunately, the whole subject of "living green" was long ago turned into a fad story and then beaten into the ground. But to people who've actually taken the pains to design an eco-home, there's nothing faddish about it; it becomes an ongoing design pursuit that is also a defining element in their lives.

Ted Baumgart, a film production designer in Los Angeles and an eco-homeowner, has so much pride in his home's design and is so eager to share its "story" that he regularly offers house tours to show off its features such as the indoor waterfall (using all recaptured water, so there's no waste), the inground plants growing *inside* the house, and the solar-powered miniature train system that runs from the kitchen to the backyard. Baumgart says, "I want to show people that creating an eco-house is very doable, and I also want to show them that, by the way, you can have a blast doing it."

Baumgart and other eco-homeowners can tell you the complete backstory behind everything in their houses—the surprising former life of those recycled windows, the weird place where the materials for the sofa originated. You may not care (though you probably should), but what matters is, they do. And that's important because, as Mau notes, if people are really going to live sustainably, and do it in a committed way as opposed to hopping on and off the green bandwagon, they have to find the meaning and the pleasure in it—and to approach it not as an obligatory gesture or dreary chore, but rather as an endlessly stimulating challenge.

Actually, the whole question of "where to live" should also be approached as a design challenge, because it's such a complex issue that has so many ramifications on your current life, your future, and on the environment. But when people are deciding where to live, they usually don't think like designers, observes Marianne Cusato. They're more like shoppers on Black Friday, anxious to grab as much home as they can. They tend to focus on the house itself and "get caught up in square footage, as well as all those checklists— 'Does it have this kind of kitchen counter and that kind of bathtub,'" Cusato says. If they applied design thinking, they'd be more apt to start by asking stupid questions (*Do I really need a big house? Why?*), then proceed to sketching out design scenarios for the short term (*How would I get around this particular area? What kind of community does it have?*) and for the long term (*Will I still want to live in this house/neighborhood ten years from now? What if my financial situation changes?*). Cusato maintains that if more people over the last decades had approached the home-buying decision in this manner, there would be less suburban sprawl and fewer McMansions. In moving farther away from cities and workplaces in order to obtain those bigger houses, many people did not consider the consequences (loss of precious time to commuting, overdependence on a car, high energy bills to heat and cool that big house, lack of a viable community or a walkable downtown). Having now experienced those consequences firsthand, some have shifted back in the direction of higher-density communities and more modest homes.

When you apply the designer's "stupid questions" methodology, a lot of assumptions about where to live and how to live start to unravel, and it becomes clear that there is much room for rethinking and redesign. Take the front lawn—it just begs the question ... *Why?* Three years ago, the designer Fritz Haeg started wondering why people devote so much acreage and energy to something so unnecessary and uninteresting. As Haeg noted, lawns must be "maniacally groomed with mowers and trimmers" which in turn causes significant levels of pollution. They use up too much precious space, drink up too much water, require chemicals that kill off everything else in sight. And with all that work, tending them is not even a creatively satisfying act, because the lawn is a celebration of "homogeneity and mindless conformity." Haeg's solution: turn front lawns everywhere into food gardens. His Edible Estates project has become a traveling exhibit that shows examples of lawn vegetable gardens as it tries to inspire others to "eat their lawns." The benefits of doing so, Haeg says, are considerable: You'll help the environment, you'll eat better, and you may even get to know your neighbors.

9.4 "WE DON'T NEED A GETAWAY—WE NEED A GET-TO"

Haeg points out that when you're growing a garden in your front yard, a funny thing happens. People notice and are inclined to stop by and ask how it's going, while you, in turn, may be inclined to give them some extra food. Social encounters of this nature can't actually be designed, but design decisions (such as the decision to grow a food garden instead of a lawn) can create conditions that lead to more social engagement.

Mau believes that achieving a greater level of engagement—with the community around you and with the issues that affect that community—is one of the keys to designing a richer, more emergent life. Here again, it's something that is often overlooked when decisions are made about what's important in life and how to allocate one's time. Research has shown that, in America at least, there's been a trend toward disengagement from community over the past two to three decades. As evidence of this, Robert Putnam, the Harvard professor and author of the book *Bowling Alone*, points to the decline of card games, civic meetings, memberships in PTAs, and, yes, bowling leagues. A much-cited Duke University study from a few years ago found that people have significantly fewer friends now than they did twenty years ago.

Design has had a hand in some of this. The design of urban sprawl, as lamented by Cusato, has been a factor; so has the design of rich and immersive home entertainment and gaming experiences. Even when design isn't encouraging us to stay home, it can still manage to cut us off—think of the "Pod people" with their trademark white earphones, who remain in their own world even while in the midst of a crowd. To a slightly lesser extent, the same is true of smartphone users. It's an irony that has been observed before, but is worth noting again: Designed technology somehow manages to connect and disconnect us at the same time.

As it becomes easier and more tempting to avoid actual, physical social engagement, it becomes necessary to design it into your life. In a culture that provides so much opportunity for escapism, Mau says, you must actively choose the "get-to" instead of the "getaway." As defined by Mau, a "get-to" is an experience that more fully engages you with the real world, with people, with matters of substance. An example of choosing a get-to over a getaway, he says, could involve your vacation: Instead of lying on a beach somewhere, "you make the decision to take your family to someplace like the Worldwatch Institute, where you can work with scientists as they record migratory patterns in the Bahamas." There are more opportunities for these kinds of hands-on experiences than ever before—in fact, there are whole new categories of travel now dedicated to more authentic cultural experiences, including hands-on volunteer-working vacations (also known as "voluntourism"). But these options are not quite as easy and automatic as more conventional vacations.

You have to seek them out, design them into your schedule. You may even have to presell the benefits—"make hope visible"—in order to rally enthusiasm among the hard-core escapists in your group.

The attempt to engage more with society is central to the "new design" approach. Mau's principle *Go deep* and the whole design industry's shift toward ethnographic research is based on the belief that design cannot evolve unless designers become more connected to and aware of what's going on in the lives of people all around them. This means that designers, like the rest of us, have a growing need to "get out more" and make new friends.

NOT SURPRISINGLY, designers try to design ways to do this, some of which are not recommended for the rest of us (for example, designers sometimes pay people to spend time with them). But there are certain techniques from design ethnography that may be more transferable. Design researchers emphasize that if you want people to allow you into their lives, and to really share their hopes and dreams with you, there is one thing you must learn to do: be a good listener. "Observing and listening are the two most important things we do," says IA Collaborative's Kathleen Brandenburg. Another thing is to try to avoid prejudging or making assumptions about people. If you approach social interactions with an open and curious mind, Brandenburg says, people will tend to surprise you—usually in a good way.

Design researchers have also discovered that social interaction between people tends to improve as you introduce elements of spontaneity and "play." IDEO has mastered the technique of setting up "play" scenarios for adults, which can involve everything from word games to making collages to playacting. Designers have relearned something that any child will tell you: Games are a great way to make friends.

There are some designers who believe that our future may be dependent on finding ways to design social interaction that, among other things, fosters the willingness and ability of adults to "play nice" with each other. Edwin Schlossberg of ESI Design in New York has been involved with this for several decades, going back to when he worked with Buckminster Fuller in 1969 on the World Game, Fuller's audacious attempt to bring people together to try to solve world problems using a game format. Schlossberg later went on to design children's museums, cultural centers, and, most recently, the fascinating new Action Center to End World Hunger in New York, where people are invited to come in off the street and join forces to try to find solutions to hunger crises halfway around the world. This and all other Schlossberg-designed environments tend to focus on "shared experiences" that encourage people to interact, play, and solve problems together.

Schlossberg maintains that as a society, we need to begin to learn (or relearn) the subtle skills of working and playing together as a group, "the ability to do something with a hundred other people." We were once better at large-group dynamics "back in the days of Fontainebleau," he notes, and some societies still are good at it. "There's an Indian word meaning 'the delight of being in large crowds,' " Schlossberg says, "but we don't have a word like that because we're so tied to the idea of individualism." Why does this matter? Because the world is growing more crowded, which makes social interaction inevitable, and also because many of the complex problems facing us are going to require group efforts in order to solve them. Schlossberg once designed a group social event during which he handed out ribbons to all in attendance and asked each person "to tie their ear to the ear of the person next to them. Then I asked them to all get up and try to leave. They were laughing hysterically because they realized how difficult it was. And my point was, 'Now you understand the idea of interdependency in complex systems.' "

The gaming world is also picking up on this notion that the future belongs to people who know how to do things collectively. Jane McGonigal, a leading designer of alternate reality games played in groups, has observed that when people play together in significant numbers, they form a collective intelligence that is capable of solving highly complex problems. And they also solve the age-old problem of finding satisfaction for themselves because, according to McGonigal, "multiplayer games are the ultimate happiness engine."

Designers are finding that a number of social problems can be addressed, in part, by figuring out ways to encourage people to interact and socialize more with one another. Hilary Cottam, who has done revolutionary work redesigning new social welfare services in the UK, has learned that when it comes to providing services for people in need—such as senior citizens, people suffering from illnesses, or low-income families—one of the keys to success is making sure that those people have a strong support network to help them with sometimes-mundane everyday problems and tasks. In short, people need a small circle of friends; and if they don't happen to have one, they may need to have it designed for them.

When Cottam's design group, Participle, began working with the London-area borough of Southwark, she was told by the local government that the rising demands of Southwark's aging population had become almost impossible to meet because "there wasn't enough money in the system to give these people all that they needed." Cottam's response was to ask the stupid question *What* do *they really need?*

After a deep dive into Southwark's senior community, Cottam found that while the government was focused on providing seniors with free car rides for errands or at-home visits from social workers (which were proving quite costly), the seniors really needed something you couldn't put a price tag on: friendship. This was important not only in terms of the intrinsic value of friendship "but also because having friends and acquaintances available can free an older person from worrying about basic chores like changing a lightbulb or getting to the store," Cottam says. It was simple, really: If someone had a few friends to count on, that person wouldn't have to be so dependent on the government.

Cottam began to devise prototypes of social circles for Southwark's seniors, with each circle consisting of a half-dozen or more people who lived near one another. The circles were not designed haphazardly; interviews were conducted to assess people's needs and capabilities, and groups were assembled for compatibility. To encourage people to join, the circle came with benefits (discounts on various services, access to group activities) funded by the government and an outside company that agreed to help sponsor the project. The program is still young, but it's already having a big impact: The buddy networks are helping members deal with health problems, household repairs, transportation, and other everyday needs that previously were overwhelming the government's capabilities. And in the process, it's reinforcing one of Cottam's core beliefs: That through better design, it's possible to improve and expand social services without spending more money.

There's also an important lesson here for a population that is aging. Community is critical at any stage of life, but never more so than in later years. Cottam maintains that, in her findings, having social connections seemed to be *the* single most important determinant of a happy old age. So if you're designing for emergence, one of the most important things to get started on is constructing that support network or "circle" that will be there for you as your life advances. The more diverse the circle, the better: Family provides one kind of support, friends another, and a mix of different ages and skill sets is important, says Cottam (she seems to have this down to a science, noting that an ideal social network as you get older "should include six people from very different roles—including a friendly professional, family member, a peer, a younger person, an older person.")

Increasingly, people are designing their own age-in-place neighborhoods (with residents chipping in to fund additional services for the neighborhood's at-home seniors), informal "Golden Girls" buddy networks, even modern-day versions of the old hippie communes, now referred to as "cohousing," where groups of people buy property together, divvy up the costs and tasks, and then promise to stick by one another for the long haul. All of these represent attempts to apply principles of design in an area where design might not seem to be needed or even appropriate; friendship is usually thought of as something that develops naturally, and as a matter of circumstance and serendipity. But designed actions and efforts can spark connections that simply wouldn't occur otherwise.

Fritz Haeg, when not attacking front lawns, has made a specialty of designing informal gatherings and get-togethers, including a longtime series of "sundown salons" that he hosted at his home, a geodesic dome in the Los Angeles area. The gatherings were organized around celebrations of art, literature, or dance; they began with people Haeg knew and soon grew to include many he didn't. He says that events such as these tend to draw people into your orbit who share similar passions and interests, "and it forms a support group that can help nurture your own creative work." Once he started doing the salons, Haeg then decided to set up an alternative school, which he called the Sundown Schoolhouse, right there in the dome. Once a week, he'd bring in an artist, scientist, or filmmaker to share knowledge with a small group, "just as a way to open the door to another way of thinking."

9.5 HOW TO REMAIN "STUPID" FOREVER

Haeg says that one of the reasons he set up his informal schoolroom was because he rejects "this idea we have that learning only happens in a college classroom." He adds, "I like the idea of learning from other people in a way that is kind of woven into your daily life, as opposed to being part of some colossal four-year education program."

It's a fortunate thing that designers such as Haeg tend to embrace "continuing education" as a way of life. They probably couldn't do their jobs well otherwise, given that design is, as Michael Bierut noted, "about everything." The nature of design requires designers to keep learning for a lifetime—about the latest technology, about sociological trends, about art and science and nature and economics, about history and pop culture, about how *everything* works, about things that no one else seems to know or care about, like the "aesthetic-usability effect" or the good old Fibonacci spiral.

So if one of the elements of designing for emergence is to keep learning for life—and Van Alstyne asserts that emergence *depends* on constant learning—then designers seem to have that down. But the rest of us may have to work at it a bit, especially since, as Haeg notes, there is a tendency for people to assume that "serious" learning happens in classrooms and not so much in everyday, postschool life. Another faulty assumption is that people are less able to learn as they get older; the old saw is that young people absorb new knowledge like a sponge, but the absorbency diminishes with age.

What some of the latest research findings show is that people do actually retain the ability to keep learning, though they absorb and process information a bit differently. Here's the interesting part: the older you get, the more you begin to think like a designer.

Recent studies cited in the 2008 edition of the neurology book *Progress in Brain Research* and subsequently reported in the *New York Times* found that as people age, their focus of attention widens and it means they can actually take in more information, not less. They notice and retain more extraneous bits of information that younger people ignore. Older minds can sometimes have trouble quickly retrieving details (like the name of that movie star, you know who I mean). However, "because they've retained all this extra data, they're now suddenly the better problem-solvers," according to Lynn Hasher, a professor at the University of Toronto. "They can transfer the information they've soaked up from one situation to another" (translation: they're *jumping fences*). Jacqui Smith, a professor at the University of Michigan, has observed that older people also are able to take new information from the situation at hand and "combine it with their greater store of general knowledge" (to form *smart recombinations*).

This dispenses with the notion that you can't keep learning forever. But how do you design your life to encourage constant learning? The answer, Mau suggests, is to intentionally and constantly "keep moving away from what you know." People tend to design their lives and careers so that they are usually on firm, familiar turf, intellectually speaking; they go with what they know. To the extent that they learn new information, it's usually in the form of deeper knowledge on a subject they're already well versed in—they increase their expertise. Mau, and many other designers, opt to go wide as well as deep. As previously noted, this is what makes designers T-shaped in their knowledge, though over time, as they continually go wide and deep, wide and deep, their knowledge base becomes more like a series of T's linked together. The horizontal line traces your continued movement into new areas of knowledge; the vertical drop-down lines represent each deep dig into a subject before you move on to the next one.

By designing your lifelong learning approach as a series of connected T's, you are building a broad base of knowledge, which is, apparently, a very Eastern thing to do, according to designer John Maeda, the president of RISD. Maeda says he first learned about this philosophical perspective at a time in his life when he was struggling to understand why he was so unfocused; he couldn't decide what to study because he wanted to study everything. "It was the kind of thing where you start to wonder, 'Am I an idiot, What's going on?'" Maeda says. "And one of my teachers in Japan said, 'No, John, it's actually quite simple. There are two ways of life. In the Eastern way, you're building a broad base, while the Western way is focused and you're building a hill straight up, and you get much higher more quickly.'" It struck Maeda that the higher, faster approach seemed more productive, and his professor said it was—in the short run. "But he explained that when you get older, the beauty of the 'broad' method is that you've built this great, grand symmetrical heap, kind of like Mount Fuji, that is solid and cannot be moved. Whereas in the Western approach, the one thing you've made keeps going up and up, but it's thin and fragile, and if one thing is wrong, the whole hill falls over."

Mau has tried to build his design firm, and his life, on a broad base of knowledge, and it can be a mixed blessing. If he specialized in one narrow area of design, he could attempt to dominate that segment and might rise up higher. Because he's so diversified and is always doing something so different from the last, it's harder for potential clients to peg him. But Mau insists he'd be miserable doing the same type of design over and over. He enjoys the feeling of being slightly "lost" in a new subject area. He likes to be the anti-expert, the naïve outsider who gets to ask the impertinent "stupid questions" and then search for the answers. Many of the people who work at his studio share this attitude. Kevin Sugden, one of BMD's senior designers, says: "We'd rather be happy amateurs than sad professionals."

THIS IS not to suggest you can't enjoy professional success being a lifelong learner and a broad-base builder. Mau has; Maeda has; and Richard Saul Wurman surely has. Wurman lives in a mansion in Newport, Rhode Island, with a massive front gate that looks like the entrance to a university. But, no, it's the entrance to Wurman's enormous house, bought and paid for by Wurman's asking and then answering a long, continuous series of stupid questions.

Wurman, in his midseventies with piercing eyes and a slightly belligerent manner, is perhaps best known as the man who created and for many years hosted the TED conferences, where people from the worlds of technology, entertainment, and design (as well as other industries) convene each year to share big ideas. Wurman is also known for several of his books, including his 1989 classic *Information Anxiety* (in which he coined the now widely used term "information architecture"). What people may not know about Wurman is that he has written more than eighty books. And while he's thought of as an expert on information design–related issues, the fact is, Wurman has written books about health care, baseball, dogs, Hawaii, Polaroid cameras, the Olympic Games, airlines, Wall Street, retirement, drugs, children, grandchildren, estate planning, and hats. Not necessarily in that order.

"Each book I do," Wurman explains, "whether it's on health care or the Olympics or on pets, is done for the same reason. It's a subject I find interesting and that I don't fully understand. And I can't find anything out there to make it understandable. So I try to make it understandable to myself. Not to others—to *myself*."

Wurman says he started the TED conferences for the same reason he wrote the books. "The goal was to take myself from not knowing to knowing. I chose technology, entertainment, and design because those were three things that interested me." But did he see the conferences as a chance to design interaction, to bring people together around big ideas? "I didn't want to bring people together," he says dismissively. "I wanted other people to pay so that I could listen to interesting people talk. I sat on the stage the whole time and they talked to me about things that I wanted to know more about."

At one time in his life, Wurman and two partners started an architecture firm that struggled for thirteen years before shutting down. He also spent a number of years as a design teacher at colleges, and he would begin classes by telling students that the big design problem wasn't designing a house or a toaster—it was designing your own life. But it took a while for Wurman to get the design right on *his* own life—he was well into his forties by the time he figured it out. His design plan was founded upon this stupid question, which he included in *Information Anxiety*: "If we are able to design our lives," Wurman wrote, "wouldn't the best result—the best measure of success ultimately—be that every day is interesting?"

Through the combination of the books and conferences, Wurman achieved the life design that answered the question and solved his problem (and it really would have become a problem, because Wurman was not suited for a regular job: "With my personality," he says, "what company would want to hire me?"). He began to emerge more fully in midlife, and he is emerging still. He says he is more curious about things today than ever before. He approaches each new subject (his latest is on the future of cities) as a kind of puzzle—what is the most important information to know, and how can it be boiled down to its most basic, most comprehensible form? Wurman's books often rely on simple charts and images to convey complex ideas. So first he learns about something that intrigues him, such as *How do we hear?*, and then, using his information design skills, he shapes the data, reduces it, clarifies it.

Wurman is not satisfied—does not feel that he truly understands something—until he can explain it, with complete clarity, to another human being. He thinks of each subject that interests him as an empty bucket; when he's reached the point of total understanding, by way of clearly explaining to others, he then considers that bucket full and moves on to the next one.

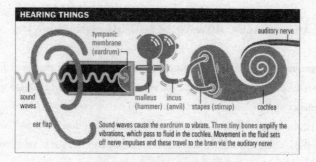

HEARING THINGS

tympanic
membrane
(eardrum)

auditory nerve

sound
waves

malleus
(hammer)

incus
(anvil)

stapes (stirrup)

cochlea

ear flap

Sound waves cause the eardrum to vibrate. Three tiny bones amplify the
vibrations, which pass to fluid in the cochlea. Movement in the fluid sets
off nerve impulses and these travel to the brain via the auditory nerve

SOUND INTENSITY

Decibels 0

10 — rustling leaves

20

30 — soft whisper, at 15 feet

40

50 — conversation, at 3 feet

60

70 — pneumatic drill, at 50 feet

80

90

100 — aircraft takeoff, at 2,000 feet

110

120 — inside a disco, at full volume

130

140

threshhold of human pain 150

160

170 — shuttle takeoff, at 2,000 feet

180

A few of the visuals on hearing from Richard Saul Wurman's Understanding Healthcare guidebook.

But what drives him to do it? "Making things understandable to yourself
and happenstantially to others is what civilization is all about—what is more
basic?" says Wurman, a bit scoldingly. "Why did they do cave paintings?
They were trying to make something understandable to themselves and others. I can't imagine a life that is devoid of that—a life where you're not trying
to come in contact with those things that you're curious about and don't
understand."

9.6 AND FINALLY: HOW TO DESIGN HAPPINESS!

His tendency to be impatient notwithstanding, Wurman is a very happy person. One reason is that he spends so much of his time at the high end of the flow channel.

According to Mihály Csíkszentmihályi, the condition of "flow" is characterized by being totally immersed and completely engaged in what you are doing, to the extent that time seems to stop. People who are in a state of flow "experience intense concentration and enjoyment, coupled with peak performance," he says. The kinds of activities that can bring us into this state are ones that involve a challenge—and there is a delicate balance involved, in that the challenge should be difficult, but also one that we "feel confident we can handle with our existing skills."

Flow occurs in the channel that separates boredom from anxiety, says Csíkszentmihályi. "As the difficulty of a challenge increases, we can become more anxious and lose flow," he explains. But you can get back into that flow, he notes, by increasing your skills so that they rise up to meet the challenge at hand. If you think of this in terms of what Wurman spends most of his time doing, he approaches each subject with very little understanding of it, which raises the "challenge" level, but he also comes in knowing that he has the skills to make that subject understandable. As soon as he masters the challenge—and begins to lose flow—he quickly moves to the next subject.

The growing body of research on happiness—which tries to define what it is and how you attain it—shows a strong connection to the state of flow. People tend to think of happiness as a goal, but it's more of a process, according to Martin Seligman, a professor at the University of Pennsylvania and the former president of the American Psychological Association. Seligman maintains that there are two activities that lead to happiness. One is what he calls the "engaging" activity—the challenging and often creative activity that tends to lead to a "flow experience."

When you're engaged in these types of creative activities, it activates an area of the brain called the nucleus accumbens that controls how we feel

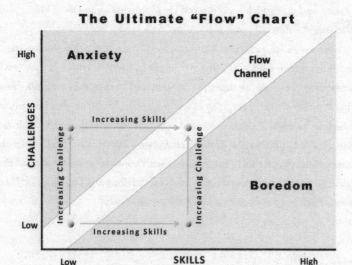

The Ultimate "Flow" Chart

about life, according to Dr. S. Ausim Azizi, chairman of the department of neurology at Temple University's School of Medicine. He noted that creative activities that you enjoy also stimulate the brain's septal zone—the "feel good" area—and that makes you feel happy.

But the other piece of the puzzle has to do with the second type of activity that can make you happy. Seligman has observed that in addition to those "engaging" or creatively stimulating activities, there are also "meaningful" activities that tend to make people happy. These, he says, involve "using what you're best at to serve others or participate in a cause bigger than yourself."

If you're doing a certain kind of design—as is Wurman, or Mau, or any "problem-solving" designer—you are combining *both* types of activities. You are creating *and* contributing to a larger cause, simultaneously. It is a recipe for, as a Chinese restaurant menu might say, double happiness.

It is also one of the keys to designing an emergent life. Through constant acts of creative design, you also re-create *yourself*. You help propel your own growth spiral, feeding off the energy of creation. That's not just a feeling, it's a fact: Being in that state of "design flow" raises the levels of neurotransmitters in your brain, such as endorphins and dopamine, and that keeps you focused and energized, according to Dr. Gabriela Corá of the Florida Neuroscience Center.

It's no wonder, then, that Mau tries to snap up all the design opportunities that come his way—all those chances to immerse himself in a newly stimulating challenge, all those portals that lead back to the nourishing flow. Mau's challenge, however, is to somehow design his life so that he ends up choosing the *right* opportunities, the ones that will offer the most possibility for growth and emergence. That's a challenge for everyone, actually, and one that the final chapter will explore, along with a few last stupid questions that might be asked by anyone starting fresh on the designed life. Namely: *What should I design? What do I do first? Where do I begin?*

10. BEGIN ANYWHERE

Why small actions are more important than big plans

10.1 NEVER MIND THE BOOKSHELF

How do you get started designing? Well, Mark Noonan's life as a designer was triggered by a heavy snowfall, which subsequently led to a backache, which in turn inspired a stupid question.

For anyone in New England, shoveling snow is a fact of life—and it was a painful one for Noonan, an investment banker living in Connecticut. After each snowfall, he invariably ended up with a sore back from shoveling, and that wasn't even the worst of it. "My wife would see me walking funny afterwards," Noonan says, "and she'd say, 'I see you've made an *S* of yourself again.'" Finally, a two-foot snowfall and another painful aftermath to shoveling fueled a resolution by Noonan in 2000 that something had to give, and it wasn't going to be his back again.

As for the stupid question, it was based on an observation. Noonan had noticed that whenever he was shoveling and felt a twinge of back pain, he'd automatically start to change the way he shoveled to take pressure off his lower back. "I'd use my knee as a lever on the shovel to lift the snow," he says. "Basically, I was making a fulcrum out of my body, and it worked." Which made Noonan wonder: *Why hasn't anybody thought of combining a shovel and a lever?* Then it occurred to Noonan that in fact somebody *had* thought of it. Now it was up to that somebody to do something about it.

Noonan set to work on his smart recombination of a lever and shovel. Along the way, a wheel came rolling into the picture. Noonan figured that

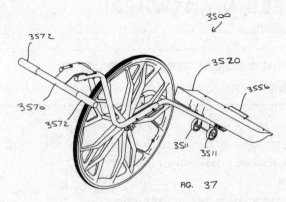

FIG. 37

while the lever lifted, the wheel could keep things moving forward (yes, he was going to try for a triple recombination). Eventually Noonan left his job so that he could work full-time on his creation. It took him two years to get the whole thing right, but when he was finished he had designed a shovel that plowed through snow twice as fast and with half the effort of a regular shovel. All the shoveler had to do was push forward and down, using body weight for leverage—no lifting involved, no strain on the lower back. Noonan christened his new design "the Wovel," pictured above in one of his patent drawings.

Today, Noonan has sold more than twenty-five thousand Wovels, and that has launched his second career as a successful product designer (he currently has dozens of different products in development and a handful already on the market, including a dog leash adaptor that magically prevents the leash from getting tangled between the dog's legs). Noonan agrees that the design process did start with asking questions about the way things work in the world and how they might be improved. And it did proceed to jumping fences, as Noonan fused together ideas and technologies from separate areas. As he progressed, he made hope visible by way of the many sketches and prototypes he created and shared with others (including his wife). In terms of doing empathic research, Noonan had it easy there because he happened to have a fine study subject in himself; you might say that through all those preceding snowstorms, whenever he trudged outside with his old flimsy shovel, he was going deep (and never more so than during that two-foot blizzard).

But Noonan says that the point at which he actually became a designer did not occur during those steps. "You could say it starts with asking questions, but the fact is a lot of people ask questions," he says. "People are always saying, 'Why doesn't somebody do this or make that,' but it doesn't go any further. It's just a rant." A person becomes a designer when he/she makes the decision to act on a problem. Says Noonan: "Instead of just asking a question, you have to take ownership of it."

NOONAN IS right: If you question the way things work but do nothing about it, you're more whiner than designer. And if you draw beautiful sketches of ideas and plans that never materialize, you may be an artist, but not a designer. While it's true that design is rooted in planning—it's "the art of planning," as Paula Scher notes—the real transformational power of design, according to the leading thinkers on the subject at Stanford University and the Rotman School, lies in the *doing*, not the planning. Designers put change in motion by acting on plans and ideas—they go ahead and try things. Each time they try, they learn, and they incorporate those lessons as they try again.

Which is why it's so important for designers to have the capacity to just . . . begin. Mau says that over the years as he has run his studio and taught at various schools, the number-one question he's been asked by young designers is, *How do I get started on designing something—where do I begin?* The question came up so much that he adopted a principle to serve as a ready answer: *Begin anywhere.*

Mau borrowed the line from one of his heroes, John Cage, who used it in the context of trying to create art or music. Cage observed that "not knowing where to begin" was a common form of paralysis for artists. So his advice was to just jump in at any point and start creating anything; start a musical composition in the middle, if need be, but start—and see where it leads. Mau was so taken with this philosophy that early in his career he designed a poster around the *Begin anywhere* theme. Through the years, the principle has remained central to his way of working, and it was part of his 1998 manifesto on the process of design.

In discussing this principle, Mau likes to tell the story of a writer he knew who was determined to write an ambitious book on a big subject. The writer, Mau recalls, "was always preparing to get started—always arranging his bookshelves, and organizing his office" so that he would have everything he needed, right where he needed it, as he began this very challenging and daunting task. But somehow, Mau says, he never did begin, and still hasn't to this day.

For designers, the temptation to "arrange the bookshelf" may be even greater than it is for writers. Design challenges are often complex, which means that one could spend endless amounts of time boning up on the subject or subjects involved. The solution may be unknown or at least unclear at the outset, which can give pause. Then there is the matter of consequences; thinking about them in advance is a good thing, but too much thought along these lines can paralyze. Mau says that designers should start putting ideas out into the world immediately, avoiding the temptation to wait "until conditions are perfect" or until all the answers to the challenge have become clear. If it's a complex design problem, those answers most likely will emerge over time and trial. In the meantime, says Mau, "you have to begin designing without all the information that you'll eventually need. The goal is to be an expert coming out, not going in."

The design consultant and author Nathan Shedroff has characterized designers as "optimistic people who are trained to be courageous about the future—and making the future happen." Even though they may be somewhat in the dark about how their efforts will ultimately play out, they plow ahead anyway, according to Shedroff. "They aren't afraid of confronting a blank piece of paper and getting to work making something new." (Indeed, Mau, Collins, and Behar seem to rarely leave any nearby sheet of paper blank for very long, taking every opportunity to fill that white space with scribbled ideas and possibilities.)

What helps designers move forward in the face of uncertainty is methodology or, to use Mau's preferred term, *process*. Given that Mau is such a big proponent of experimental thinking and creative "wilding," one might expect he'd be resistant to using a formalized design process, but the opposite is true. Mau thinks that process enables experimentation. "It's like a safety net," he says. People tend to feel more comfortable experimenting with new ideas and venturing onto unfamiliar turf when they carry with them an established method of working and solving problems. It means that even if they don't quite know what they're doing, they always know what to do.

The process Mau uses has been loosely described throughout this book. He often starts by questioning standard practices; then begins an experimental-thinking phase that is described in Chapter 2 as jumping fences; he quickly begins to create sketches, prototypes, and other representations of ideas that can be shared; and at some point, he brings empathic research into the mix.

Mau likes to do these things in a certain order (for instance, he strongly believes that research should follow initial creative brainstorming). But another way of thinking about the design process holds that the order of doing certain things doesn't matter; it only matters that you do those things at some point in the process.

That's the view of George Kembel, recognized as one of the foremost experts on design methodology as practiced by IDEO and other cutting-edge design firms. Kembel has been a driving force, along with IDEO's David Kelley, behind the formation of Stanford University's groundbreaking design program, the d.school, which is attempting to turn grad students into leaders who can transform industry and tackle social problems by using design principles. The school's students come from a variety of academic backgrounds, often with very little design training. And so one of Kembel's main tasks is to figure out, in his words, "What does it take to shape students to go out in the world feeling confident they can tackle big design challenges they haven't seen before?"

10.2 GOING 'ROUND IN CIRCLES AND FILLING UP BUCKETS

The key to creating confident designers is to give them a reliable process. Once the students have internalized a certain way of looking at and then attacking problems, *voilà*—they're ready to design a solution to just about anything, Kembel maintains. The basics of the design process that Kembel teaches are surprisingly simple (it's the mastering of the techniques that is hard).

And in fact the Stanford process as taught by Kembel is not all that distinct from Mau's, but it's broken down and presented slightly differently. Kembel points out that while there are any number of "three-step, and seven-step, and ten-step processes used by different designers," he tends to boil it all down to five basic requirements, not necessarily in any order: 1) gaining expertise about a problem or subject area, primarily through empathy with the people directly involved; 2) framing the challenge you're going to tackle (which is to say, making sure you're asking and trying to answer the right questions); 3) generating options or ideas; 4) creating prototypes to test those options; and 5) iterating, or creating subsequent refined versions of your original prototype, based upon feedback.

Those are the building blocks of good design and it doesn't really matter which one you start with (in other words, you can begin anywhere) as long as you circle around to each one, and do so repeatedly. Kembel suggests thinking of each of these requirements as an empty tank or bucket, with the five of them placed in a circle: "As you go through the design process, you're filling up each tank a little bit more while you go around and around. And by the end you have a lot of expertise, a lot of empathy for the problem, confidence that you've framed the right challenge, tons of ideas generated, and lots of testing so you know what's working."

It's a basic methodology that can work for a professional in a design firm or for an individual trying to solve almost any type of design challenge, according to Kembel. For the individual, it may require a bit more resource-fulness at some of the stages. For example, when it comes to empathic re-

search, a firm like IDEO has client budgets and an infrastructure in place for doing deep dives, as well as highly trained observational research specialists such as Jane Fulton Suri. But on some level, Kembel says, anyone can gain design insights from observation. Kembel's students do it all the time. They simply venture out into the world and ask people to allow them to tag along and watch.

This can lead the students into all kinds of eye-opening situations. Once, the d.school students followed and observed firefighters to try to figure out what type of communication device might help them do their jobs better. The students learned that when firefighters are about to enter a burning building, they don't know the exact locations of the worst hot spots, the areas most in danger of imminent collapse. This sparked an idea among the students that was wildly inventive and interesting enough to spur further ongoing research by the Motorola company. It involved designing miniature disposable transmitters that the firefighters could toss, like marbles, into a burning building, so that the devices could relay back localized temperature information to the firefighters' handheld radios.

Most empathic research has a lower thrill rating than that—more along the lines of watching people as they do chores or ride the bus. The idea, says Kembel, is to "engage with people in everyday life" and to do so with antennae raised at all times. He trains students to use IDEO's Five Whys method (shown in Chapter 1) and to reflexively say things like, "That's interesting— tell me more." Sometimes not saying anything at all is the most important thing an empathic researcher can do; the idea is to leave plenty of openings for other people to talk. Kembel also teaches would-be ethnographers to look in "adjacent areas" for research insights. For example, when working on a project about aging and mobility, students didn't just visit senior facilities, they also studied young athletes on campus who'd been injured, to see how they dealt with limited mobility.

With regard to the "framing" stage—comparable to the *ask stupid questions* phase in the Mau model—the Wovel designer Mark Noonan offers this useful tip: "Try to keep refining your question, to make it more specific. "You start with 'How do I shovel snow without hurting my back?' and gradually the question evolves to something like, 'How do I move snow more efficiently by shifting the weight?' If you keep refining the question in this way, sometimes the solution will become self-evident."

As ideas do start to take shape, it's critical to prototype them. Prototyping is a form of play, really; it's a return to drawing pictures and building Lego figures or paper airplanes. Artistic skills are not necessary, but a shift in the way you think about work may be required. "As adults, most of us are not used to showing our ideas in very rough, unfinished form," Kembel says. "We've been taught through the years that you must present ideas in a highly polished, 'professional' manner." But designers know that the earlier and more often you get ideas in front of other people, the better. Even a rough sketch on a napkin or a mock-up pasted together with Elmer's glue can serve as a vehicle that goes out into the world and returns with critical feedback, which is particularly valuable if it comes from people with firsthand knowledge of the problem or situation being addressed. Tim Brown of IDEO talks about "building to learn": Only by making a series of prototypes does IDEO learn what it's doing wrong at each stage, and that learning is incorporated into the next version of the prototype.

One of the greatest mistakes novice designers make, according to Kembel, is to hold back their ideas instead of sharing sketches and prototypes with others. "People tend to be too protective of their ideas, too early," he says. "If innovation equals creativity plus execution, most people seem to shy away from the second part." The reluctance to share ideas may be based on fear that ideas will be stolen, but applying for patent protection can ease some of those concerns (for more information on design patents, see the Resources section). More often, though, people are just leery of subjecting an idea to criticism. Kembel advises, "If you've got a good idea, it's best just to get it out in front of other people—because if it can't survive that, then you may not have something worth spending your time on."

Kembel says he's been surprised to see that as students learn the design methodology, "there's an awakening that happens—first, they see how these principles of empathy and prototyping apply to a few projects, and then it becomes a way of looking at the world as they start to apply it to other parts of life. It goes from being a methodology to a mind-set."

When that happens, he says, students are apt to apply the design methodology to solving problems around the house or in the garage. They may apply it to career planning or organizing events with friends. All of which is fine by Kembel. He thinks design can solve big problems in the world, but he also believes that "there's nothing wrong with starting small, and close to home."

10.3 WAKING UP TO A NEW DAWN, IN THE AGE OF "AQUARIASS"

Often, in fact, that's where design innovation happens: with a person trying to address a simple everyday need in life. A number of the "basement Buckys" interviewed for this book were people who started out by getting caught up in a desire to solve one simple problem. Prior to that, they tended not to think of themselves as designers and certainly not as inventors. Those two terms are sometimes used interchangeably to describe people who create new and useful things, but the word inventor can be more intimidating because it seems to suggest that one must invent from scratch, as opposed to remaking and improving things that already exist in the world. The word also has a bit of a "mad scientist" connotation to some: "I hate the term because it just doesn't seem practical," Wovel's Noonan says.

For would-be designers trying to figure out how or where to get started, the answer may be staring them in the face. That was the case with Gauri Nanda, a college student in Boston who had a basic problem with the alarm clock on her bedside table: It simply wasn't effective in getting her out of bed in the morning. "I would hit the snooze bar repeatedly," Nanda says. "Some of my friends who had the same problem said they tried putting their alarm clock on the other side of the room. But because they knew where it was, they would just sleepwalk to it, turn it off, and go right back to bed."

Nanda's offbeat idea was to design an alarm clock that would wake you up and then flee, all the while beeping, forcing you to wake up enough to give chase. She put wheels on her clock but also had to design a landing system with shock absorption so that when the clock rolled off the night table and dropped to the floor, it could survive the fall, right itself, and keep rolling and beeping. An alarm clock that taunts and evades its owner in this manner had better be cute, Gauri felt, so she also sought to design a clock with a more personable face, as her earliest sketches show.

scared/anxious

big lips

circle mouth/round head

The original prototype for the Clocky, as Nanda called it, consisted of "little more than a pair of Lego wheels and motors and a shag-carpet covering to hold it together." The actual creation of the product took about a year as it went through a series of iterations. The big challenge was to make it light and small, but still durable enough

box/small face

to survive daily dives off a three-foot-high night table. The Clocky drew a good deal of media attention when Nanda introduced it a few years ago, eventually showing up for sale at the MoMA design store, and its success has helped her to launch her own product design firm.

Nanda isn't the only person who is trying to design a better way to wake up in the morning. Ian Walton and Eoin McNally, a couple of students at the Royal Society of Art in London, went in a whole different direction from Nanda. If the Clocky is designed to jar you awake, their "glo Pillow" aims to do the job gently and quietly, using nothing but light. Walton says that he started on the project by "going deep" into sleep research. He learned that a good way to wake from sleep is by being exposed to a source of light that starts dim and then gradually increases—as is the case when you awaken to a sunrise (this is another reminder that you often find the best design solutions in nature, as biomimicry expert Janine Benyus points out in Chapter 8). Walton and McNally tried to mimic the natural rousing effects of a sunrise by embedding a grid of LED lights into a programmable pillow. The idea of putting lights into a pillow started as a joke, Walton says. It's such an obvious idea that Walton and McNally laughed it off at first but then came back to it. The pillow itself offered a way to soften the light while also putting that light, quite literally, in your face. The key was to design in subtlety; once the pillow is programmed, a very faint light begins to glow less than an hour before wake-up, and gradually the pillow emits more and more light until it's as bright as the morning sun.

If Nanda, Walton, and McNally have shown that you can design your way out of bed in the morning, other designers have been working on what happens after that. Todd Greene reinvented his morning shave with a radically different razor; Jay Sorensen found a way to avoid getting burned by his morning coffee; Eleanor Leinen developed a better pair of shoes to take her out the door and into the world. A designer looking for challenges to work on could simply chart the moments and experiences that comprise an entire day and find design opportunities at every turn.

Some of these everyday-need designers approached a problem from a universal standpoint: Sorensen's Java Jacket coffee cup sleeve is something every coffee drinker needs—which is why that corrugated paper cup sleeve can now be found in takeout coffee shops around the world. The idea came to Sorensen in 1991 after he spilled burning coffee in his lap while trying to hold a hot cardboard cup with a slippery napkin wrapped around it. The Java Jacket may look simple, but it took more than a few design iterations to come up with this ubiquitous product.

On the other hand, some are designing more for their own particular needs. Greene created his popular HeadBlade razor to make life easier for men who shave their pates. His glimmer moment came, believe it or not, while he was rubbing his head deep in thought about something else. He says that it suddenly occurred to him: "If I made a razor that fit under my fingers, I could shave just by running my hand over my head." He started sketching designs in his notebook.

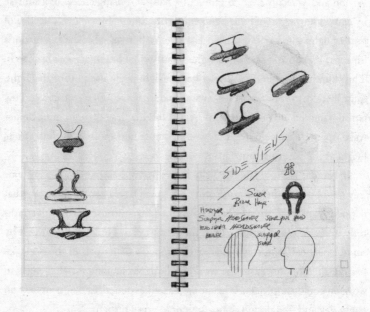

As for Leinen's shoes, they were designed to meet the special needs of diabetics like herself. The only shoes on the market for people with her condition "weren't just ugly—they were an abomination," she says. So Leinen proceeded to design colorful shoes that offered the special fit and comfort needed by diabetics, without skimping on style.

Though most of them weren't schooled in design processes and principles, the do-it-yourself designers instinctively followed the same steps as the students of Kembel or Mau—framing and reframing the question or challenge at hand, undertaking various forms of empathic research, and working through multiple iterations with homemade prototypes. Greene, for example, says he tried a dozen HeadBlade prototypes before he got it right. His test models were nothing fancy: "I went to my local crafts store for some clay, then stopped at the drugstores for razor handles and blades," he says. "I would sculpt a design and put it in the oven until it was hard." He showed his prototypes to others and tried them out on his own willing head. When a prototype didn't work, he broke it apart and used the pieces to begin again.

WHAT DRIVES these people to design is not just need but desire—the desire "to come up with a creative solution to a problem that no one else has been able to solve," says Steve Greenberg, who has written about Greene and other designer/inventors in his book *Gadget Nation*. Greenberg has observed that when novice designers do succeed in solving a problem, something changes in them, too: They become more observant of design shortcomings all around them but see these failings as potential opportunities to create something better. And as they continue to design—even if it's only in the basement, after work—"they end up reinventing themselves," Greenberg says. They start to think of themselves as designers or inventors in a way that supersedes whatever else they do—"it doesn't matter if they're plumbers, chiropractors, whatever," he says.

Then again, they might be pop stars or moguls. The drive to design seems to be found in people from all walks, including those who could easily afford to pay others to solve design problems for them. Presumably, Justin Timberlake could enlist the best golf course designer in the world on his behalf, but Timberlake has recently opted to try to design his own course.

Brad Pitt has taken to designing houses. Academy Award winner Daniel Day-Lewis designs and makes his own wood furniture.

The musician and artist David Byrne designs, among other things, funky chairs and bicycle racks. Why chairs? Because, Byrne says, "chairs are anthropomorphic by nature—they have legs, arms, and are in some ways mirror images of people." When Byrne sketches a chair, like the Organic Pads chair below, he says he feels as if he is creating "portraits, possibly self-portraits." And why bike racks? Simple: Byrne has a passion for bicycling, and his home city of New York doesn't offer enough places to secure a bike. When Byrne was approached by city officials to help with the problem, he sketched some designs, and "they said if I could get them fabricated, they'd put them up. When opportunity knocked, I opened the door." He's designed a number of fanciful site-specific racks for various city neighborhoods, including a curvaceous female figure for "Olde" Times Square.

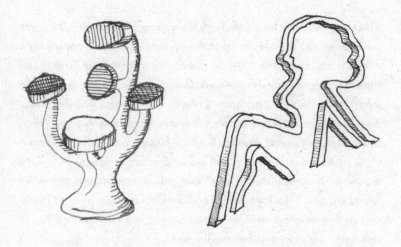

Celebrities sometimes engage in these kinds of highly personal design activities without anyone even knowing about it. When Google launched its patents search engine a couple of years ago, a design blogger searched through the database and discovered that famous people were quietly designing and making all kinds of quirky things. Among the finds were detailed diagrams revealing actress Jamie Lee Curtis's quest to design the perfect diaper (with a built-in moisture-proof pocket for holding cleanup wipes), as well as rock guitarist Eddie Van Halen's endeavor to make a "musical instrument support" to prop up his guitar during two-handed solos, and even an elaborate attempt by Gary Burghoff (who played the character Radar on *M·A·S·H*) to engineer something he called "an advanced fish attractor device."

The blogs had some snarky fun with the discovery that famous people were hatching plans to improve the way we live, but it's somehow refreshing to find that movie stars and rock idols can turn out to be basement Buckys, too. When the words *celebrity* and *design* are combined, what we typically get is another of those ubiquitous star fragrances or clothing lines, which seem to have more to do with leveraging a popular name than satisfying a genuine need (and it's usually not clear if the star involved did more than attach his/her name to a product). But the Curtis diaper or the Burghoff fish attractor represents a more authentic kind of design—born of a passion to solve a problem and evidence of long hours of scheming and sketching in private (until the Google patents site made it all public).

Some of the stars' dream products may seem a bit humble, but as Kembel has suggested, humble is not a bad place to start with design. The legions that have joined the rapidly growing craft movement aren't out to remake the world so much as to dress up and beautify their own tiny corner of it. Meanwhile, their techie counterparts, the gadget makers, also are often pursuing a narrow and personal vision, trying to solve problems that in many cases, frankly, no one asked them to solve. (Dog-powered scooter, anyone?) In the current "golden age for gadgets," people are scouring the Internet for information on how to build things and then using their own Web sites to try to sell those creations. Hence, everywhere you look these days, there are new attempts at smart recombinations—though they don't always get the "smart" part right. For example, it turns out there was a very limited market for, the MP3 Pez dispenser, a Frankensteinish attempt by one designer to put a Pez head on an iPod body. Slightly more successful but no less weird is designer Oliver Beckert's hybrid that turns a bathroom toilet tank into a fish aquarium (and if that seems a bit tacky, well, so is the name: Beckert has called it the "Aquariass").

It's safe to say the world doesn't urgently need a toilet aquarium or an MP3 Pez dispenser, though there's certainly room for playful creations that aim to elicit smiles. Did the world have any inkling how much it needed the Frisbee before it was designed? But for those aspiring designers inclined to try to create something more useful and worthwhile, particularly in serious times of great need, the question arises: What *does* the world need now? When this question was posed to Kembel, his response was: "What *doesn't* it need? Is there any area you can think of that *couldn't* benefit from better thinking and design?"

10.4   WATER, WATER EVERYWHERE? THINK AGAIN.

As to which areas of need are most pressing, it depends on whom you ask. Several of the designers featured in this book are focused on water issues, which would seem to indicate two things: It's a gathering concern, and it's an area that offers many points of entry for designers. As previously noted, Project H's Emily Pilloton is trying to design ways to help people transport water, while both Dean Kamen and LifeStraw's Mikkel Vestergaard Frandsen have developed better ways to purify it. The makers of the Aquaduct bike have tried to design a way to both move it *and* clean it. Brian Collins, meanwhile, is focusing on visualization of the problem; given that most people don't even know that there *is* an issue with water, Collins believes that well-designed awareness campaigns are a critical starting point. Mau is focused on getting companies to cut down on water waste. And on top of all that, there are designers working on better ways to recapture rainwater, or to pull water from the air, or even to get restaurants to stop overserving water and then throwing it away. There are infinite ways to come at this one complex, multifaceted problem.

The same could be said of one of Hilary Cottam's specialty areas: providing a better life for older people. Cottam thinks designers need to focus on the fact that as our society is rapidly aging, we don't have the tools and services in place to meet that challenge, and government doesn't seem to be up to the task. While Cottam has been devising better ways to deliver services to seniors in the UK, primarily by designing social support networks, other designers are exploring ways to re-create the conventional retirement community so that it doesn't isolate seniors from the rest of the community and other age groups. There's also a growing demand for "universal design" housing, household tools, and furniture made to accommodate people with limited physical capabilities—as well as a need for labels and road signs that can be read by people with compromised vision. And, as the design writer Bruce Nussbaum has pointed out, in this age of social networking it is an unfortunate reality that seniors—who need companionship and "connection" as much as anyone—tend to be excluded from the digital community. One British designer, Ben Arent, is already taking on this challenge with a prototype version of a Facebook-type network for seniors called Jive, designed to be simpler and more accessible for low-tech users. That's one of countless design possibilities in this area—because, in fact, virtually anything that can be designed can also be redesigned with senior needs in mind.

On the issue of transportation, there's clearly a continuing dire need for better car design (Gordon Murray's city car, described in Chapter 8, is just one of many design approaches being tried), along with an even more critical need for redesigned public transit systems that can help lure people out of their cars. But if reinventing cars or subway systems seems too daunting, how about focusing on the humble bicycle? Increased bike usage can lessen traffic congestion while also reducing pollution and helping riders stay fit, which means there are a number of good reasons to try to design a better bike—and lots of people are doing that, churning out new iterations that are lighter, more foldable, and easier to ride for people of all ages (IDEO recently designed a "coaster" bike for Shimano that is proving popular among aging boomers who like to, well, coast). There are smart recombination bikes out there, taking on all kinds of problems and chores; Toronto's Niki Dun designed one that doubles as an ambulance, with attached medical equipment that folds out.

Actually, however, the most important design advances may be happening in the ecosystem of biking. For example, the city of Portland, Oregon, is using design to cut down on the incidence of bike riders getting hit by cars. By taking a Jane Fulton Suri–like analytical approach to recent accidents, Portland officials figured out that bike lanes weren't sufficient to protect bikers from "right hook" collisions, which happen when a car's driver is making a right turn and can't see the bike rider alongside on the right. The design solution: creating "bike boxes" at busy intersections, which provide a more visible area, positioned out in front of cars, where the cyclist can wait for the traffic light to change. Meanwhile, other cities around the world, most notably Paris, have tried to encourage more riding by designing innovative bike-sharing programs, with on-street locations where you can pick up and return bikes anytime, for a minimal charge. In Boulder, Colorado, Alex Bogusky—he of the Mini car launch and the teen antismoking Truth campaign—is trying to devise a similar system, but he envisions creating bike-share stations with computer kiosks enabling people to keep track of how many miles they've ridden and calories they've burned while also doing some online social networking with other bikers in the area.

For those who want to help design solutions for people in need of shelter, Cameron Sinclair, out there on the front lines, is always looking for new ideas and helping hands; information on connecting with Architecture for Humanity is in the Resources section. But you don't have to journey to the far reaches of the world; lots of people are designing solutions for the homeless in local areas, as Tom Fisher's Minneapolis students have done, and as Steve Mykolyn did with his 15 Below coat.

If your interest is in helping to redesign education, you can zero in on one of the many tools of learning—as Yves Behar is doing with the XO-2 laptop—or you can help redesign the school itself and the ecosystem around it. After-school programs, school nutrition, mentoring: Every program in every school system requires design, and the success of those programs can be dependent on how much planning, empathy, and prototyping has been brought to the effort.

There are many more possibilities that can be explored on the Web. For example, Brian Collins's Designism group, mentioned in Chapter 7, has launched a site that pairs up designers with various social sector projects, working through the Idealist.org volunteer site. And if you log on to the Web site of the InnoCentive network, you can click through the current design challenges that are up for grabs, both in the business and the nonprofit sectors. If you already have a world-fixing design project in the works and want to shine a light on it, there are a number of international design competitions that focus on life-improving design, including the Denmark-based INDEX: AWARD (this show recently awarded prizes for a bottle that uses solar energy to clean water and a small first aid tool called the Tongue Sucker, which enables the user to help an unconscious person to breathe by opening air passages). And in addition to award shows, there are now cash prize contests encouraging designers to solve real-world problems. The recent Automotive X Prize contest, aimed at encouraging the design of a super-fuel-efficient car, is probably the best known, but it's just one of many. Details on how to connect with some of these organizations and competitions can be found in the Resources section.

WHEN IT comes to moving designers to take action, it seems the best source of inspiration is life experience. Most innovative projects do not originate with a designer doing a Web search or entering a competition. They tend to begin with someone venturing out into the world, looking around, and noticing a problem or need. Think of Dean Kamen, going out to buy himself an ice cream and seeing a man in the rain, struggling to get over the curb with his wheelchair; or Jock Brandis, who was moved to create his nut-shelling device after seeing up close how painful and difficult it was for African women who had to shell peanuts by hand.

Kamen and Brandis didn't go out looking for problems to solve. They simply paid attention to what was happening around them. And they also did another thing that all designers do, or should do: They empathized. For a moment, they put themselves in the shoes of someone else and observed that the shoes needed improvement. Blake Mycoskie did likewise, though what he noticed was that the shoes were missing altogether.

Mycoskie's is another of those great accidental design stories, whereupon an outsider wanders into a strange place, starts looking around and asking stupid questions, and then starts connecting ideas to bring about change. In this case, the place was an impoverished village in Argentina, where twenty-nine-year-old Mycoskie had gone to do some volunteer work while taking a break from his job in Los Angeles. Mycoskie's observation—it was pretty hard to miss—was that the kids in this village all lacked shoes. And he wondered: *How could I change that situation?* He then jumped the fence by connecting something seemingly insignificant that he'd noticed earlier during his trip—that urban Argentinians tend to wear inexpensive and quite comfortable canvas slip-on shoes called *alpargatas*—with the needs of those kids, and soon he landed on a big idea. He would take the concept of those plain canvas *alpargatas* back to the United States, spiff up the design a bit with colored prints, and then market them with a promise: For every pair of the shoes sold, Mycoskie would donate a pair to the kids in Argentina.

Mycoskie created a company, TOMS Shoes (the name refers to "tomorrow's shoes"), with a business model that could have been dictated by Bruce Mau. Everything the company did was designed around a big, bold initiative of giving shoes to children in need. Mycoskie didn't have to advertise to get noticed; he proved Mau's point that a company can design its actions to speak louder than ads. TOMS Shoes became a hot media story, written up in a slew of magazines and newspapers, and an even hotter product. Mycoskie ended up with a successful business along with a National Design Award, while ten thousand kids in Argentina ended up with shoes, handed out by Mycoskie and his co-workers in 2007. He has since done similar giveaways of hundreds of thousands of shoes in the United States, Ethiopia, and South America.

Mycoskie says he believes the best ideas tend to be found in unexpected places, when you're not even looking for them. "I think you just have to put yourself out there in the world," he says, "and be open to what's going on around you."

10.5 THE ORIGINAL MANIFESTO, REVISITED

Mycoskie's point about "putting yourself out there" echoes something Mau wrote in his original manifesto: *Take field trips*, he urged, adding, "the bandwidth of the world is greater than that of your TV set, or the Internet, or even a totally immersive, interactive, dynamically rendered, object-oriented, real-time, computer graphic–simulated environment."

It's a good lesson, and one of many embedded in that original document Mau wrote more than a decade ago and which still gets quoted and passed around the Internet every day. Designers are always discovering it anew, posting it on their blogs, telling people to "check out numbers 5, 13, and 27," and so forth. Five of the "laws" in it became principles in this book (*Ask stupid questions*, *Jump fences*, *Go deep*, *Work the metaphor*, and *Begin anywhere*), and a number of others have been mentioned in various chapters.

The manifesto was written in a weekend, "as a favor to my wife's sister," Mau recalls. His sister-in-law was launching a new magazine and needed material. "I was busy with my own work, but she was relentless." Once he started writing, he says, "My ambition was to set out the patterns in the wild organic processes of innovation. I tried to articulate the way that we work—day to day, moment by moment—so that others might learn from the method we had developed."

After giving it to his sister-in-law, Mau then brought the manifesto with him to a design/tech conference in Amsterdam, where he was one of the guest speakers. The conference was all about the latest software applications, and so everyone was surprised when Mau read the part of the manifesto that declared that "creativity is not device-dependent." He urged the audience to put aside technology every once in a while and just "think with your mind." The techies loved it.

Since then, Mau says, "No other product of the studio has had the distribution and resonance of the manifesto. It seems to have a life of its own." It's been translated into Russian and German and adopted by various organizations, including, Mau likes to point out, a pro tennis association.

Whether it will help with a serve or backhand is questionable, but if you're embarking on a life of design, or really any creative endeavor, here are a few rules from the manifesto worth bearing in mind:

Forget about good. Good is a known quantity. Good is what we all agree on. Growth is not necessarily good. Growth is an exploration of unlit recesses that may or may not yield to our research. As long as you stick to good you'll never have real growth. Mau thinks it is particularly important for people who want to innovate to disregard what other people think is "good." Those who've brought about real change, he notes, "weren't focused on making good design, they were concerned with making history."

Study. Use the necessity of production as an excuse to study. Mau has always believed that a design studio should be a place of study and that designing should be an exercise in lifelong learning. As Richard Saul Wurman showed, if you can somehow bring together work and learning—"using the necessity of production as an excuse to study," as Mau puts it—you're at least partway on the road to designing a richer, more emergent life. With that in mind, Mau recommends making your own design studio, wherever it may be, into an environment that encourages learning. Surround yourself with ideas; stock the place with books. Just don't spend too much time arranging the bookshelf.

Imitate. Don't be shy about it. Try to get as close as you can. You'll never get all the way, and the separation might be truly remarkable. There is something quite liberating about this law. It serves as a reminder that designers, and creators in general, needn't feel they must reinvent the wheel. They can start just by trying to make their own version of a wheel; along the way, hopefully, they will add something new to that wheel (a shovel? an alarm clock?), and then it will become an original design. Meantime, who needs the pressure of being an "inventor"—of being the next Edison? It can stop you from ever getting started. There are few completely original ideas; just about everything you think of has been influenced by something else. But if you can connect "old" ideas in new ways, you can create progress.

Slow down. Desynchronize from standard time frames and surprising opportunities may present themselves. To put it another way, design thinking takes time. To get into the flow, you must separate yourself from flux. It's no easy thing: Much of daily life these days is spent responding to endless stimuli—the constant stream of phone calls, meetings, e-mails, tweets—and we tend to accept and even welcome the interruptions because, as Stefan Sagmeister has noted, it's easier to react than to create. But those who can, at times, screen out the distractions and focus their attention on one thing, observes journalist David Brooks, "have the ability to hold a subject or problem in their mind long enough to see it anew."

To create an environment that lends itself to this kind of deep thinking, some people need an austere "quiet room," while others opt for something more playful (Brian Collins recommends turning a space into your own personal kindergarten classroom, with chalkboards and walls covered with drawings and other scraps of inspiration). The décor may not matter as much as the wiring, or the desired lack thereof. Too many interruptions can disrupt the connections and recombinations that may be forming in the designer's mind. One study, by Hewlett-Packard, found that constant interruptions actually sap intelligence (by about ten IQ points, in fact). To truly disconnect and desynchronize, Sagmeister takes periodic and lengthy creative sabbaticals to remote places: at the time of this writing, he was somewhere in the Far East, unreachable except by hand-delivered letter. But whether one journeys to another country or another part of the house, the goal is to end up, in Mau's words, "lost in the woods" with a creative challenge or a problem that needs solving.

Allow events to change you. _You have to be willing to grow. Growth is different from something that happens to you. You produce it. You live it. The prerequisites for growth: the openness to experience events and the willingness to be changed by them._

MAU HAS allowed events to change him. A couple of months after his trip to Florida for personal redesign, he was slimmed down and clean shaven. He claimed to have more energy—but then Mau has never lacked for energy.

His new life design plan seemed to be working. He was using his time much more efficiently, and that was a good thing because his work schedule seemed jammed. Mau was working with old clients MTV, Coke, Indigo stores, and Shaw carpeting, while taking on new projects such as the Mideast city redesign. He figured out how to do it without shuttling back and forth constantly on planes. "I think part of redesigning my life has been accepting that I can't do so many things myself—but the collective energy of the studio can," he says. "I just have to be more willing to say to someone, 'I trust you to drive this project.'"

Meanwhile, stateside, he'd begun to roll out a couple of new iterations of his Massive Change idea. One involved a new program designed to take the idea far and wide by way of a network of schools, or "centers for massive change." This was a plan brought to Mau by one of his new partners, Seth Goldenberg, who suggested it might be possible to effectively franchise the concept of massive change to universities or companies, enabling them to set up their own design/innovation labs using Mau's methodologies and his brand. As part of the program, Mau also wants to make Massive Change courses available online, so that anyone—from a grad student to "the kid in India"—can have access to the problem-solving tools of design thinking. It's a radical idea and still in its early stages, but if Mau can pull it off, he might be able to tap into the growing hunger out there among people wanting to design solutions to something—anything.

This ambitious project was being formulated by Mau in the midst of a very cold winter in America, with the recession biting hard on everyone, including the design world. Mau was sober about the realities. He knows that some of his projects are quite vulnerable. Design is often the first thing to be cut by those who continue to think of it as a superficiality.

Nevertheless, Mau was brimming with optimism. He viewed the recession as one more important wake-up call. Mau thinks—and a lot of experts share this view—that hard times are, in some sense, the best time for design. With constraints everywhere, designers are needed to negotiate them, sometimes overcome them, and sometimes help us figure out how to live with them. And with optimism becoming a more scarce and precious resource, designers must play the role of making sure that hope remains visible, touchable, shareable.

Mau says his work on the "In Good We Trust" project, which is taking stock of volunteer and charitable initiatives as well as all manner of innovations, has convinced him of this much: The spirit of wanting to contribute, of trying to solve problems and make the world better, is stronger than most people realize or understand. "*Something* is happening out there," Mau states. "People are seeing the world around them with greater resolution, because there is more information available to them. For the first time, they are seeing patterns that have been invisible to the larger public in the past. We are moving toward global clarity—to people finally being able to say, 'Oh, I get it. Now I see what's going on in the world.'"

And what happens when everyone sees and understands and recognizes the need for change? If you ask Mau, what happens next is—of course— design.

the glimmerati

DEBORAH ADLER (*Chapters 1, 3*) designed the ClearRx prescription medicine packaging system used by Target stores, the first major redesign of that type of packaging in the past forty years. She has worked as a designer at Milton Glaser Inc. and recently launched her own independent design operation, focused primarily on designing for health care needs.

PAOLA ANTONELLI (*Chapter 8*) is a senior curator in the Department of Architecture and Design at the Museum of Modern Art in New York. As curator of influential shows such as MoMA's Design and the Elastic Mind, Antonelli is attempting to promote a wider understanding of design's influence.

IRENE AU (*Chapters 5, 6*) is director of user experience at Google, where her team is responsible for design and user research for Google's products worldwide. Prior to Google, she spent eight years at Yahoo!, where she was vice president of user experience and design.

SHIGERU BAN (*Chapter 8*) is an architect and designer known for his use of eco-friendly and economical materials such as cardboard and paper, which he has used to construct, among other things, emergency shelters in Rwanda and a church in Japan. Ban has studios in Tokyo and New York.

YVES BEHAR (*Intro; Chapters 1, 2, 4, 5, 6, 8, 10*) is an industrial designer and founder of fuseproject, the San Francisco–based firm he established in 1999. He is the chief industrial designer of One Laptop Per Child's XO laptop computer, and is working on a sequel. Other clients include Birkenstock, Bluetooth, Mini, and Herman Miller.

JANINE BENYUS (*Chapter 8*) is a pioneer in the field of biomimicry, which studies natural design and tries to extract lessons that can be applied to man-made design. Benyus is the author of *Biomimicry: Innovation Inspired by Nature*. Her company, the Biomimicry Guild, offers biological consulting, and her Web site AskNature.org answers questions about natural design.

JOHN BIELENBERG (*Chapter 7*) is a designer, educator, and founder of C2 Design in San Francisco and the Bielenberg Institute at the Edge of the Earth in Belfast, Maine. He also directs Project M, a program that strives to inspire young creative individuals so that their work can have a significant and meaningful impact on the world.

MICHAEL BIERUT (*Chapters 1, 2, 3, 4, 9*) is a graphic designer, design critic, and a partner in the New York office of Pentagram. Bierut is one of the founding editors of the influential blog Design Observer. His work is represented in the permanent collections at the Museum of Modern Art in New York and at the Cooper-Hewitt.

ALEX BOGUSKY (*Intro; Chapters 5, 10*) is a partner and creative director at Crispin Porter + Bogusky, an ad agency with offices in Miami and Boulder. CP+B is responsible for a number of innovative campaigns that have fused advertising and design, including work on the teen antismoking campaign "Truth" and the U.S. launch of the Mini Cooper.

KATHLEEN BRANDENBURG (*Chapters 4, 9*) is director of design strategy and cofounder, along with creative director **Dan Kraemer,** of the Chicago-based IA Collaborative design firm. The firm's clients include Nike, Hewlett-Packard, Caterpillar, Charles Schwab, Adobe, and the Nutrition Rich Foods Coalition, for whom IA Collaborative is currently developing a new way of presenting nutritional information designed to help people eat more healthily and judiciously.

TIM BROWN (*Chapters 1, 3, 4, 5, 10*) is the CEO and president of the design firm IDEO. He speaks, writes, and blogs regularly on the subject of "design thinking" and its relationship to innovation in business. Brown also has a special interest in the ways design can be used to promote the well-being of people living in emerging economies.

BILL BUXTON (*Chapters 3, 6*) is principal researcher at Microsoft Research and the author of *Sketching User Experiences: Getting the Design Right and the Right Design*. Previously, he ran the Toronto-based firm Buxton Design, was a researcher at Xerox PARC, and was chief scientist of Alias Research and SGI Inc.

VALERIE CASEY (*Chapters 1, 6, 8*) heads a global practice at IDEO, where she designs socially and environmentally sustainable products, services, and business models for organizations around the world. In late 2007 Casey founded the Designers Accord, a call to arms for the creative community to reduce the negative environmental impact caused by design.

LEE CLOW (*Chapter 6*) is chief creative officer at TBWA Worldwide and has overseen many of the marketing campaigns for Apple and other top brands. Clow believes that marketers must move beyond ads to actually design the ways they behave, as evidenced in the cultural redesign of the Pedigree dog food company.

BRIAN COLLINS (*Intro; Chapters 1, 3, 5, 7, 9, 10*) headed the Brand Innovation Group at Ogilvy & Mather before launching his own experiential branding firm, COLLINS:, which is doing work for Al Gore's Alliance for Climate Protection. Collins also teaches design at the School of Visual Arts and oversees the annual forum Designism: Design for Social Change, sponsored by the Art Directors Club of New York.

HILARY COTTAM (*Chapters 8, 9, 10*) is a founding partner of the London-based firm Participle, which is attempting to redesign social services in the UK by developing fresh approaches to education, caring for seniors, and other services. Cottam has also worked on the redesign of prisons. In 2005 she was named UK Designer of the Year.

MARIANNE CUSATO (*Chapters 7, 9*) is the designer of the "Katrina Cottage," conceived in 2005 as an alternative to the FEMA emergency trailers supplied to survivors of Hurricane Katrina. The Katrina Cottage won a National Design Award in 2006. Cusato, based in Miami, runs a design firm that bears her name.

MARTIN FISHER (*Chapter 8*) is the cofounder of KickStart, which has developed innovative water pumps and other low-cost tools aimed at increasing productivity of small farmers in Africa. More than fifty thousand new microenterprises have been started using KickStart equipment. Fisher continues to direct a team of designers and engineers in Kenya.

HEATHER FRASER (*Chapters 4, 6*) is director of Designworks, a center for design-based innovation and education at the University of Toronto's Rotman School of Management. Fraser helped develop the Three Gears of Design model that is featured in this book.

FRANK GEHRY (*Intro; Chapters 2, 6*) is a Pritzker Prize–winning architect whose best-known works include the Guggenheim Museum in Bilbao, Spain, IAC headquarters in New York, and the Walt Disney Concert Hall in Los Angeles. He has collaborated with Bruce Mau on a number of projects, including the Disney Concert Hall and the soon-to-be-open Museum of Biodiversity in Panama.

MILTON GLASER (*Chapters 1, 2, 3, 6, 7, 9*) is among the most celebrated graphic designers in the United States and a recipient of the lifetime achievement award of the Cooper-Hewitt National Design Museum. As a Fulbright scholar, Glaser studied with the painter Giorgio Morandi in Bologna. He opened Milton Glaser, Inc. in 1974 and is an articulate spokesman for the ethical practice of design.

MICHAEL GRAVES (*Intro; Chapters 1, 5*) has been at the center of the democratization of design owing to his successful collaboration with Target stores. Graves is famous for designing stylish teakettles and other household objects, but he has also been an influential architect since the 1960s, known for designing the interiors of his buildings down to the furniture and lighting fixtures. In recent years, Graves has begun to focus on "universal design" for the physically impaired.

BOB GREENBERG (*Chapter 5*) is founder of R/GA, a leader in interactive marketing design. His firm helped design the NikePlus system, which serves as a model of how a product can be elevated to an experience that helps build a community. Greenberg is an outspoken critic of conventional advertising and advocates for a new marketing model with design and technology at the center.

FRITZ HAEG (*Chapter 9*) divides his time between his architecture and design practice Fritz Haeg Studio, the ecology initiatives of Gardenlab (including Edible Estates), and other various combinations of building, designing, gardening, exhibiting, dancing, organizing, and talking. In 2006 he initiated Sundown Schoolhouse, the self-organized educational environment based in his geodesic dome in Los Angeles.

ALEXANDER ISLEY (*Intro; Chapter 1*) runs a Connecticut-based design studio bearing his name and specializing in brand development and communication design for organizations involved with culture, fashion, and retail. Isley first gained recognition in the early 1980s as the senior designer at Tibor Kalman's influential M&Co. He also served as the first full-time art director of *Spy* magazine.

DEAN KAMEN (*Intro; Chapters 1, 2, 3, 7, 8, 10*) is an inventor and designer who holds more than four hundred U.S. patents. His company, DEKA, has produced the Segway, the iBOT walking wheelchair, and the Slingshot water purification system. Kamen also founded FIRST (For Inspiration and Recognition of Science and Technology), which sponsors science competitions for high school students.

LARRY KEELEY (*Chapters 4, 5*) is an innovation strategist and cofounder of the Doblin Inc. business consultancy. He is a board member of IIT's Institute of Design. Keeley is known for deconstructing different kinds of innovation and for creating three-dimensional "innovation landscapes" in presentations to clients. He also developed the Compelling Experience Framework featured in this book.

DAVID KELLEY (*Chapters 4, 10*) was an unhappy electrical engineer when he enrolled in Stanford University's design program. He subsequently founded a design firm in 1978 that became IDEO (Greek for "idea"), now with 400-plus employees worldwide. He has helped design the first computer mouse, the Palm Treo, and the Leap chair. Kelley also has taught design at Stanford for the past twenty-five years.

GEORGE KEMBEL (*Chapters 6, 8, 10*) is cofounder and currently executive director of Stanford University's d.school, which has emerged as a leader in the teaching of "design thinking." Kembel heads up a graduate program that teaches a highly structured design methodology to business, engineering, and design students.

REM KOOLHAAS (*Intro; Chapter 2*) is a world-renowned architect who heads up the Office for Metropolitan Architecture (OMA), which has completed, among other projects, the Dutch embassy in Berlin, the Prada Epicenter in Los Angeles, and the Seattle Public Library, where Koolhaas collaborated with Bruce Mau. In 1995, he and Mau published *S, M, L, XL*, a book that documents the work of OMA.

GEORGE LOIS (*Chapters 1, 2*) is an award-winning art director and designer and one of the original "Mad Men." Known for the striking graphic covers for *Esquire* magazine that he designed during the 1960s, Lois also was a partner in several ad agencies and was a pioneer of the "creative revolution" in American advertising. His breakthrough ad campaigns included "I Want My MTV" for a then-struggling MTV.

JOHN MAEDA (*Chapters 1, 3, 7, 9*) is president of the Rhode Island School of Design. A renowned graphic designer and computer scientist, Maeda was originally a software engineering student at the Massachusetts Institute of Technology (MIT), where he eventually completed a PhD in design and went on to teach. He is the author of *The Laws of Simplicity*.

ROGER MARTIN (*Intro; Chapter 2*) is dean of the Rotman School of Management in Toronto. Named one of the ten most influential business professors in the world by *BusinessWeek* in 2007, Martin writes extensively on design and innovation, and has a special interest in "integrative thinking," which he explores at length in his book *The Opposable Mind*.

STEVE McCALLION (*Chapter 5*) is creative director of Ziba Design in Portland, Oregon, which has been a pioneer in using design research for major companies. McCallion launched Ziba's consumer insights and trends group, using a blend of ethnography and "cool hunting" to anticipate what people are looking for next. His work for Umpqua Bank is featured in this book.

WILLIAM McDONOUGH (*Chapter 8*) is the founding principal of William McDonough + Partners, a design firm practicing ecologically, socially, and economically intelligent architecture, and is also principal of MBDC, a product development firm assisting companies in designing eco-friendly solutions. McDonough and partner Michael Braungart coauthored the groundbreaking *Cradle to Cradle: Remaking the Way We Make Things*.

CLEMENT MOK (*Chapters 1, 3*) is a designer, digital pioneer, software developer, and former creative director at Apple who helped design its early graphic interfaces. Since then, Mok has founded several successful design-related businesses: Studio Archetype, CMCD, and NetObjects. He has served as chief creative officer of Sapient and as the president of the AIGA design group.

GORDON MURRAY (*Chapters 2, 8, 10*) is a renowned designer of Formula One race cars and the famous McLaren F1 "supercar," though currently his design firm (Gordon Murray Design Ltd in Surrey, England) is focused on creating the prototype T.25 city car—which is smaller than a Smart car and incorporates a cradle-to-cradle design featuring flexible architecture and a reusable body and chassis.

DONALD NORMAN (*Chapters 1, 4, 5*) is a preeminent author on the subject of design whose influential books include *The Design of Everyday Things* and *The Design of Future Things*. A cognitive scientist who first became interested in design while researching the Three Mile Island disaster, Norman is a veteran of Apple and now runs the Nielsen Norman Group consulting firm. He is also a professor at the University of California and at Northwestern University.

BRUCE NUSSBAUM (*Chapters 2, 4, 10*) has been a leading voice on the growing role of design in business while serving as an editor and writer at *BusinessWeek* magazine. His blog, Nussbaum On Design, focuses on innovation and design thinking. In 2005 he was named one of the forty most powerful people in design by *I.D. Magazine*. Nussbaum is also a professor of innovation and design at The New School.

VAN PHILLIPS (*Chapter 2*) lost his foot in a water-skiing accident in 1976 and spent the next two decades designing the Flex-Foot, a high-performance carbon composite prosthetic foot manufactured by the Ossur company and sold under the name Cheetah. Currently, more than 90 percent of Paralympian athletes use Cheetahs. Phillips also founded the Second Wind Foundation to help amputees by providing inexpensive and virtually indestructible prostheses.

EMILY PILLOTON (*Chapters 7, 10*) is founder and executive director of Project H Design, a San Francisco–based charitable organization focused on socially conscious design initiatives for "Humanity, Habitats, Health, and Happiness." Trained in architecture and industrial design, Pilloton lectures internationally about design activism and is the author of *Design Revolution: 100 Products That Empower People*.

STEFAN SAGMEISTER (*Chapters 1, 2, 10*) studied design in Vienna and launched his New York–based design firm, Sagmeister Inc., in 1993. He has done branding, graphics, and packaging for clients as diverse as HBO and the Guggenheim Museum, and also has designed iconic album covers for Lou Reed, the Rolling Stones, David Byrne, and Aerosmith. Sagmeister's most recent book is *Things I Have Learned in My Life So Far*.

PAULA SCHER (*Chapters 1, 2, 9, 10*) has designed corporate identities, posters, environmental graphics, packaging, magazines, public spaces, and just about everything imaginable. Her images, including those created for the Public Theater, have come to be visually identified with the cultural life of New York City. A partner at Pentagram since 1991, Scher is a member of the Art Directors Club Hall of Fame and was awarded the profession's highest honor, the American Institute of Graphic Arts (AIGA) medal.

EDWIN SCHLOSSBERG (*Chapters 7, 9*) was a protégé of Buckminster Fuller and worked on Fuller's groundbreaking 1969 World Game. Schlossberg went on to found ESI Design, a leader in experiential design. He designed the Brooklyn Children's Museum, one of the first interactive learning environments, along with other innovative museums, parks, retail environments, and public spaces, including, most recently, the Action Center to End World Hunger.

CAMERON SINCLAIR (*Intro; Chapters 7, 8, 10*) is the cofounder, along with Kate Stohr, of Architecture for Humanity, a charitable organization based in San Francisco that seeks design solutions to humanitarian crises. Sinclair and Stohr recently launched the Open Architecture Network, the world's first open-source community dedicated to improving living conditions through innovative and sustainable design.

PHILIPPE STARCK (*Intro*) is a French product designer whose work ranges from lavish interiors to fashionable household items and furniture (his Louis Ghost chair is a design icon). Starck is associated with high-end, stylized design, but recently, in announcing plans to focus more on socially responsible projects such as designing windmills, he declared: "Everything I designed is absolutely unnecessary."

DAVIN STOWELL (*Chapters 1, 4*) cofounded Smart Design with partner Tom Dair in 1980 after he and Dair graduated from Syracuse University. Early on, the firm was approached by an entrepreneur seeking to make a potato peeler that his arthritic wife could easily use; hundreds of prototypes later, the OXO Good Grips peeler emerged. Smart Design has gone on to help create many other innovative products, including the Flip camcorder.

JANE FULTON SURI (*Intro; Chapters 4, 10*) is partner and creative director at IDEO. Early in her career, she used her background in psychology to help the British government as it tried to determine why people were injuring themselves using certain products. She then became a pioneer in bringing psychology-based research into the field of industrial design. She is the author of *Thoughtless Acts?*, about the subtle and amusing ways in which people interact with the world.

JOHN THACKARA (*Chapters 1, 2, 7, 8, 9*) is a former London bus driver who is now focused on driving social change by way of design. A self-described "symposiarch" (someone who designs and produces collaborative events, projects, and organizations), he served as director of the Netherlands Design Institute and also founded the international design conference Doors of Perception. In his book *In the Bubble*, Thackara coined the phrase "smart recombinations."

GREG VAN ALSTYNE (*Chapters 3, 4, 9*) joined Bruce Mau Design in its start-up days and later headed up Mau's Institute Without Boundaries educational program and Massive Change exhibit. He also founded MoMA's Department of New Media. Today he heads the Strategic Innovation Lab at the Ontario College of Art & Design, where he has done extensive research and writing (with coauthor Robert K. Logan) on the subject of "designing for emergence."

MASSIMO VIGNELLI (*Chapters 1, 2, 3, 7*) is a world-renowned designer whose work has ranged from corporate identity, package, and furniture design to the creation of a version of New York City's subway map that is no longer used but is still revered. He is a recipient of the Lifetime Achievement Award from the National Museum of Design at Cooper-Hewitt. Born in Milan, Italy, he runs his New York–based firm, Vignelli Associates, with his wife and business partner, Lella.

PATRICK WHITNEY (*Chapters 4, 6*), director of the Institute of Design at Illinois Institute of Technology, is known as a prominent advocate of human-centered design. Whitney has published and lectured throughout the world about how to make technological innovations more humane, as well as on the link between design and business strategy. In Chapter 4, he explains his concept pertaining to the "innovation gap" (and how to close it).

RICHARD SAUL WURMAN (*Chapters 3, 9, 10*), architect, graphic designer, and author, is considered a pioneer in making complex information clear. His eighty-plus books include *Information Anxiety* and the award-winning *Access* travel guides. He coined the term "information architecture," a notion that holds that the explosion of data requires systemic design to make it understandable. Wurman also is the creator of the highly influential TED conferences, uniting the themes of technology, entertainment, and design.

GIANFRANCO ZACCAI (*Chapter 6*) is cofounder and president/CEO of Continuum, an innovation consultancy that uses design research to identify compelling business opportunities. Continuum has played a key role in the development of breakthrough products such as Procter & Gamble's Swiffer. Zaccai and Continuum are also dedicated to exploring the power of design in relation to developing nations.

the glimmer glossary

ABDUCTIVE REASONING An approach to problem-solving and creative thinking that looks beyond *what is* and speculates on *what could be.* (*Chapter 2*)

AFFORDANCE Refers to the fundamental properties of a thing "that determine just how the thing could possibly be used," writes author Donald Norman. An object's affordance suggests or invites interacting with it a certain way; a chair affords sitting, while a bowling ball affords a three-finger grip. (*Chapter 2*)

BIOMIMICRY is "the study of natural design in order to extract lessons and principles that can be applied to man-made design," per author Janine Benyus. (*Chapter 8*)

BUTT BRUSH Paco Underhill, a specialist in observing behavior in stores, discovered that if a shopper was brushed from behind by someone passing down the aisle, often the shopper would put down whatever he/she was looking at and abandon that location. Underhill's observation led to store redesigns that incorporate wider aisles. (*Chapter 4*)

CAD MONKEY A term used in the book by Cameron Sinclair to describe a low-level designer who does repetitive work on a computer (such as designing the same machine part over and over). CAD refers to computer-aided design, which relies on software that can be used to visualize and refine designs through digital simulations. (*Chapter 7*)

CHUNKING The technique of separating information into "chunks" of conceptually related content and arranging those chunks by giving precedence to critical information while deemphasizing what's less important. (*Chapter 3*)

COMPELLING EXPERIENCE FRAMEWORK An attempt by the innovation consultancy Doblin Inc. to deconstruct a human consumer experience into a clearly defined set of stages: attraction, entry, engagement, exit, extension. (*Chapter 5*)

CONSTRAINTS The term can refer to restrictions or barriers that designers themselves build in to devices or systems to prevent people from doing the wrong thing (such as a door that only opens toward you and keeps you from blindsiding people on the other side). But *constraints* is also used more broadly to refer to the parameters, restrictions, and requirements inherent in a particular design challenge. (*Chapter 8*)

CRADLE-TO-CRADLE A term popularized by the sustainable design guru William McDonough, who maintains that designers up till now have been making objects destined for the "grave"—meaning a landfill or incinerator—when they could and should be designing things that return to the cradle, reborn in new forms. (*Chapter 8*)

DEEP DIVE Refers to designers' use of extensive, in-depth research, often done in the research subjects' natural habitat rather than in an interview room or focus group setting. Designers doing a deep dive may temporarily move into people's homes, eat meals with the family, and "shadow" people on daily rounds of shopping, work, or errands, for days or weeks at a time. (*Chapter 4*)

DESIGN THINKING A process that endeavors to solve problems and create new possibilities, generally by relying on empathic research (studying people to try to figure out what they need) combined with creative experimentation and extensive prototyping and refinement—all aimed at the goal of producing better, more useful objects, experiences, services, and systems. (*Chapters, 1, 4, 10*)

DINOSAUR BABY A term coined by IDEO designer Paul Bennett to describe a quirky and idiosyncratic design creation that is destined to be loved only by its creator. (*Chapter 2*)

DISTRIBUTED POSSIBILITY Per Bruce Mau, this refers to today's widespread dissemination of design tools, useful knowledge, and expanded capabilities. Solo designers can go online and learn about a problem, find out what has and hasn't been tried, download technical data, connect with experts, seek out collaborative partners, and show off prototypes on YouTube. (*Chapter 7*)

EMERGENCE The way simple organisms and communities grow and evolve over time, often in complex, unexpected ways. The nature of emergence is such that it cannot be fully controlled or designed, but it is possible to arrange and design conditions that encourage it. (*Chapter 9*)

EMPATHIC RESEARCH Observational research that focuses on studying people to try to uncover and understand their latent needs and wants. (*Chapter 4*)

EMPATHY TOOLS Aids and tools to help researchers and designers empathize with the people they are designing for. For example, designers creating products for arthritics or people with impaired vision may wear heavy gloves or fogged eyeglasses as empathy tools during the design process. (*Chapter 4*)

EPHEMERALIZATION A term coined by Buckminster Fuller, who believed that ongoing advances in technology, if properly harnessed and utilized, could provide the opportunity for designers to "do more with less"—to achieve more functionality and affordability in designs, while using less energy, less materials, and generating less waste. (*Chapter 8*)

EXPERIENCE DESIGN The practice of trying to orchestrate all the variables likely to inform the overall experience of using a product or service. In moving from designing objects to designing experiences, designers "must step back and envision a long sequence of events someone goes through as they interact with your design," explains Smart Design's Tom Dair. (*Chapter 5*)

EXPLORATORY SKETCHING Designers sometimes use the process of sketching not just to depict and share ideas, but also to find them. Designer Milton Glaser describes the process as "a conversation between the hand and the brain [that] results in the development of an idea." (*Chapter 3*)

EXTREME USERS People who are either extremely familiar with or completely unfamiliar with the type of product or service being designed. For a toothbrush designer, an extreme user could be someone who brushes obsessively or someone who is toothless. The former will tend to find creative new ways to use existing products, while the latter is apt to make mistakes during use; either one is instructive to a designer. (*Chapter 4*)

FEATURITIS A condition that results when designers try to cram too many features or functions into a device, making it overcomplicated and inadvertently making it harder to use. The phenomenon is also referred to as "feature creep." (*Chapters 5, 9*)

FIBONACCI SPIRAL An organizing principle of nature. As certain organisms in nature emerge, such as sunflower heads, pinecones, and snail shells, they follow an efficient spiral growth pattern that has the same structural properties at any scale. This pattern corresponds to a mathematical formula known as the Fibonacci sequence in which each number is the sum of the preceding two numbers. (*Chapter 9*)

FORGIVENESS Per the book *Universal Principles of Design*, forgiveness in design "helps prevent errors before they occur, and minimizes the negative consequences of errors when they do occur." For example, in designing a laptop for children, Yves Behar had to expect that kids would push the wrong buttons and might even occasionally drop the computer on the ground—so the design had to anticipate and "forgive" such actions. (*Chapters 1, 8*)

FRACTURE CRITICAL A term that describes systems and objects that were not built to withstand a single part failure—such as when a weak gusset leads to a bridge collapse or one failed bank sends a financial system into chaos. The term comes from Thomas Fisher, who heads up the University of Minnesota design program. (*Chapter 7*)

FRAMING A technique used by designers to establish the parameters of a situation or problem and also to emphasize what's most important while clarifying the desired objectives. In effect, the frame becomes the agreed-upon "box that you will be working in as you try to solve a problem," says Ziba Design's Steve McCallion. In a separate context, designers (and other communicators) also use framing to manipulate the way information is presented to an audience, by highlighting some information and downplaying other information. (*Chapter 1*)

HEURISTIC BIAS A trained way of doing things, in which we tend to repeat the same learned behavior without thinking about it. Designers take into account the heuristic biases of others when designing, and at the same time designers may also take pains to overcome their own heuristic biases (the techniques and approaches they've used many times before), so that they can explore new ways of thinking and creating. (*Intro; Chapter 2*)

IMMERSION "A state of mental focus so intense that awareness of the 'real' world is lost," according to psychologist Mihály Csíkszentmihályi. Generally, a person who is immersed in an experience also loses track of time. Experience designers trying to create an immersive environment (such as at Umpqua Bank) must achieve a delicate balance of providing a sense of comfort and relaxation along with challenge and stimulation. (*Chapter 5*)

INNOVATION GAP The widening chasm between the knowledge of how to make things as compared to the knowledge of how people live and what they actually want/need. According to the Institute of Design's Patrick Whitney, this has led to a situation in the business world in which companies know how to make anything—but have no idea what they should make. (*Chapter 4*)

INTEGRATIVE THINKING Refers to "the ability to hold two opposing ideas in the mind at once, and then reach a synthesis that contains elements of both ideas but improves on each," according to Roger Martin of the Rotman School of Management. (*Chapter 2*)

ITERATION In design, and particularly in complex design, success is likely to arrive in stages, via a series of experiments or iterations. The "design thinking" model of problem-solving involves creating many iterations, in rapid succession, so that the designer can learn from doing and from feedback at each stage. (*Intro; Chapters 2, 10*)

LATERAL THINKING A term coined by educator Edward de Bono regarding the solution of problems through an indirect and nonlinear approach. Designers intent on finding fresh solutions and ideas may try to "think laterally"—sometimes by intentionally pursuing random thoughts and forging illogical mental connections—in an effort to veer off the path of familiar thoughts. (*Chapter 2*)

LAWS OF SIMPLICITY According to John Maeda, president of the Rhode Island School of Design and author of *The Laws of Simplicity*, there are ten principles that can help in designing for simplicity, the most important being: "Simplicity is about subtracting the obvious, and adding the meaningful." (*Chapter 3*)

MAPPING In terms of product design, mapping refers to the intuitive relationship between controls and their effects; if you turn a wheel left and the car moves left, that's an example of good mapping. Used in another sense, design researchers employ *mapping* to track and chart the way people think and act in given situations. For example, IDEO uses *behavioral mapping* to create intricate diagrams showing patterns of physical movement among people being studied. (*Chapters 2, 4, 5*)

MASS-TO-INFORMATION RATIO Designers often face the challenge of having to present a maximum amount of useful information about an object without increasing the "mass" (the size of the object, the amount of space for labels and instructions, etc.), according to the designer Brian Collins. Referring to Deborah Adler's redesigned prescription bottles, he notes that, "Without changing the bottle's mass much, she managed to put in ten times as much information—she completely shifted the [mass-to-information] ratio." (*Chapter 3*)

PERSONAS Fictitious characters created to represent the different types of people that might be expected to use a designed product, service, or space; personas help designers to remain focused on the needs of the end user when making design choices and decisions. (*Chapter 4*)

PROTOTYPING The process of sketching, building, or in some way producing a series of rough, unfinished versions of a design. Prototyping is important because it enables designers to learn as they build, based on testing and feedback. (*Chapter 3*)

SATISFICING Settling for what is good enough instead of pushing on toward an optimal design solution that might be too difficult or costly to achieve. (*Chapter 8*)

SHADOWING A research step that involves closely following people in their daily lives and routines for research and design insights. In China, Nokia's researchers shadowed users in dark apartment hallways—and saw that people were using their cell phone screens to illuminate their way down the hall. The next generation of phones for China was designed with a flashlight feature. (*Chapter 4*)

SLOW DESIGN An offshoot of other "slow" movements, this is a trend toward designing in a more thoughtful way that takes into account factors such as the effect of design on the environment and on local communities. The term also refers to designed objects and experiences that invite contemplation, "mindfulness," and sharing. (*Chapter 9*)

SMART RECOMBINATIONS Designer and author John Thackara has used this term to refer to mental connections that are actually problem-solving insights. A smart recombination connects existing ideas and possibilities that would seem to be unrelated, resulting in a new idea. (*Chapters 2, 3, 9, 10*)

T-SHAPED PEOPLE Refers to people who have a deep interest and expertise in one skill—that's the vertical base of the T—but who also, over the course of their careers, continue to branch out into many different areas of knowledge. IDEO's Tim Brown maintains that the best designers are T-shaped people and that, ideally, as they evolve, both the base and the top of the T should keep growing together. (*Chapter 1*)

THINKERING A blend of *thinking* and *tinkering*, this refers to a process in which working with physical (and sometimes virtual) objects can serve to stimulate productive inquiry, learning, and new ideas. The phenomenon of thinkering is currently being studied by IIT's Institute for Design (which has begun to experiment with creating "thinkering spaces" in libraries). The term's origins are unclear, but one early appearance was in Michael Ondaatje's novel *The English Patient* where it was used to suggest "collecting a thought as one tinkers with a half-completed bicycle." (*Chapter 3*)

THREE GEARS OF DESIGN According to Heather Fraser of the Rotman School of Management, observational research to try to better understand user needs is the first gear in what she calls the Three Gears of Design for business. The second gear takes those observations accumulated in that first phase and uses them as the basis for creating sketches and prototypes of new ideas and innovations. The third gear involves redesigning a company's systems of activities (the way it operates) in order to realize and optimize the possibilities that emerge from those first two gears of the design process. (*Chapters 4, 6*)

TOUCH POINTS Refers to all the various points of contact between a brand and consumers—packaging, advertising, in-store experience, telephone product support, etc. To design a cohesive experience, all the touch points should be integrated. (*Chapter 5*)

TRANSFORMATION DESIGN An attempt to completely reinvent existing structures, systems, and organizations by applying principles of design. In the business sector, for example, transformation design can be used to cohesively change many aspects of a company to meet demands for more openness, innovation, transparency, sustainability, consistency, and corporate responsibility. Bruce Mau uses the term "massive change" to refer to this kind of cultural/organizational transformation. (*Chapters 1, 6*)

TRANSFORMATIONAL METAPHOR Wherein a company might begin to think of itself as something very different, which, in turn, would change the way it behaves (for example, a bank might begin to model itself after a boutique hotel). In his 1998 manifesto, Mau declared that every product, every service, every brand "has the ability to stand for something else." When a company takes this approach, it can have a liberating effect by opening up new possibilities and offering fresh ways to present itself to the public. (*Chapter 5*)

UNFOCUS GROUP To try to move beyond the rigidity of question-and-answer focus groups, IDEO researchers began to conduct "unfocus groups" in which participants were encouraged to tell stories, draw pictures, and assemble collages—all designed to elicit contributions from the whole group and release inhibitions. Meanwhile, snippets of side conversations between group members were recorded, to get a sense of what people were *really* thinking. (*Chapter 4*)

UNIVERSAL DESIGN A form of design that aims to achieve accessibility for all end users, including those who may be aging, have disabilities, or have other special needs. To succeed, this type of design must be flexible, simple, intuitive, perceptible (easily understood), and forgiving, according to the Center for Universal Design. (*Chapter 10*)

WAYFINDING Refers to the process by which people attempt to navigate through an experience or a physical space relying on available cues and information as a guide. Designers may try to present and arrange that information in a manner that leads people along a desired path. (*Chapter 5*)

WICKED PROBLEMS Multifaceted and complex problems whose incomplete or contradictory nature is such that each attempted solution often seems to create a new problem. (*Chapters 7, 8*)

For more definitions of design principles, I recommend the book Universal Principles of Design *(Rockport, 2003), which expands on one hundred common design concepts with illustrations and examples.*

resources

BOOKS

A Whole New Mind: Why Right-Brainers Will Rule the Future, Daniel H. Pink, Riverhead Books, 2005

Art of Innovation, The: Lessons in Creativity from IDEO, America's Leading Design Firm, Tom Kelley with Jonathan Littman, Currency/Doubleday, 2001

Artscience: Creativity in the Post-Google Generation, David Edwards, Harvard University Press, 2008

Brand Apart: Insights on the Art of Creating a Distinctive Brand Voice, Joe Duffy, One Club Publishing, 2005

By Design: Why There Are No Locks On the Bathroom Doors in the Hotel Louis XIV and Other Object Lessons, Ralph Caplan, St. Martin's Press, 1982; new edition, Fairchild Books, 2005

Change by Design: How Design Thinking Can Transform Organizations and Inspire Innovation, Tim Brown, HarperBusiness, 2009

Cradle to Cradle: Remaking the Way We Make Things, William McDonough and Michael Braungart, North Point Press, 2002

D.I.Y.: Design It Yourself, Ellen Lupton (editor), Princeton Architectural Press, 2006

Design and the Elastic Mind, Paola Antonelli (editor), MoMA, 2008

Design Encyclopedia, The, Mel Byars, MoMa, 1994, 2004

Design for the Other 90%, Cynthia Smith, Cooper-Hewitt National Design Museum, 2007

Design for the Real World: Human Ecology and Social Change (second edition), Victor Papanek, Academy Chicago Publishers, 1984

Design Is . . . : Words, Things, People, Buildings, and Places, Akiko Busch (editor), Princeton Architectural Press, 2001; Metropolis Books, 2002

Design Life Now, National Design Triennial, Barbara Bloemink, Brooke Hodge, Ellen Lupton, Matilda McQuaid, Cooper-Hewitt National Design Museum, 2007

Design Like You Give a Damn: Architectural Responses to Humanitarian Crises, Cameron Sinclair and Kate Stohr, Metropolis Books, 2006

Design Literacy: Understanding Graphic Design (second edition), Steven Heller, Allworth Press, 2004

Design of Everyday Things, The, Donald A. Norman, Basic Books, 1988, 2002

Design of Future Things, The, Donald A. Norman, Basic Books, 2007

Designing Brand Experiences: Creating Powerful Integrated Brand Solutions, Robin Landa, Thomson/Delmar Learning, 2005

Do You Matter?: How Great Design Will Make People Love Your Company, Robert Brunner and Stewart Emery with Russ Hall, Pearson Education/FT Press, 2008

Edible Estates: Attack on the Front Lawn, Fritz Haeg, Metropolis Books, 2007

Emotional Design: Why We Love (or Hate) Everyday Things, Donald A. Norman, Basic Books, 2005

Flow: The Classic Work on How to Achieve Happiness (second edition), Mihály Csíkszentmihályi, Rider, 2002

Fortune at the Bottom of the Pyramid, The: Eradicating Poverty Through Profits, C. K. Prahalad, Wharton School Publishing, 2006

From Edison to iPod: Protect Your Ideas and Make Money, Frederick W. Mostert and Lawrence E. Apolzon, Dorling Kindersley, 2007

Gadget Nation: A Journey Through the Eccentric World of Invention, Steve Greenberg, Sterling, 2008

How to Think Like a Great Graphic Designer, Debbie Millman, Allworth Press, 2007

In the Bubble: Designing in a Complex World, John Thackara, MIT Press, 2005

Information Anxiety 2, Richard Saul Wurman, QUE, 2001

Innovation Killer, The: How What We Know Limits What We Can Imagine . . . and What Smart Companies Are Doing About It, Cynthia Barton Rabe, AMACOM, 2006

Laws of Simplicity, The: Design, Technology, Business, Life, John Maeda, MIT Press, 2006

Life Style, Bruce Mau, Phaidon Press, 2000

Long Tail, The: Why the Future of Business Is Selling Less of More, Chris Anderson, Hyperion, 2006

Make It Bigger, Paula Scher, Princeton Architectural Press, 2002; Springer, 2005

Massive Change, Bruce Mau and the Institute without Boundaries, Phaidon Press, 2004

Michael Graves Designs: The Art of the Everyday Object, Phil Patton with Michael Graves Design Group, Melcher Media, 2004

Objects of Desire: Design and Society Since 1750, Adrian Forty, Cameron Books, 1986; second edition, Thames & Hudson, 1992

Opposable Mind, The: How Successful Leaders Win Through Integrative Thinking, Roger Martin, Harvard Business School Press, 2007

Seventy-nine Short Essays on Design, Michael Bierut, Princeton Architectural Press, 2007

Sketching User Experiences: Getting the Design Right and the Right Design, Bill Buxton, Morgan Kaufmann, 2007

Subject to Change: Creating Great Products and Services for an Uncertain World, Adaptive Path: Peter Merholz, Todd Wilkens, Brandon Schauer, David Verba, O'Reilly Media, 2008

Substance of Style, The: How the Rise of Aesthetic Value Is Remaking Commerce, Culture & Consciousness, Virginia Postrel, HarperCollins, 2003

Things I Have Learned in My Life So Far, Stefan Sagmeister, Harry N. Abrams, 2008

Thoughtless Acts: Observations on Intuitive Design, Jane Fulton Suri, Chronicle Books, 2005

Toothpicks & Logos: Design in Everyday Life, John Heskett, Oxford University Press, 2002

Understanding Healthcare, Richard Saul Wurman, TOP, 2004

Universal Principles of Design: 100 Ways to Enhance Usability, Influence Perception, Increase Appeal, Make Better Design Decisions, and Teach Through Design, William Lidwell, Kritina Holden, Jill Butler, Rockport Publishers, 2003

Visual Thinking for Design, Colin Ware, Morgan Kaufmann, 2008

Wired to Care: How Companies Prosper When They Create Widespread Empathy, Dev Patnaik with Peter Mortensen, FT Press, 2009

WEB SITES, ORGANIZATIONS, & COMPETITIONS

AIGA
http://www.aiga.org
The Web site for the professional association for design.

Architecture for Humanity
http://www.architectureforhumanity.org
Cameron Sinclair's volunteer nonprofit organization set up to promote architecture and design solutions to global, social, and humanitarian crises.

Bamboo Bike Project
http://www.bamboobike.org
A project by scientists and engineers at the Earth Institute at Columbia University that aims to examine the feasibility of implementing cargo bikes made of bamboo as a sustainable form of transportation in Africa.

Before & After, Inc.
http://www.before-after.com/index2.html
Tom Monahan's creative thinking site.

The Buckminster Fuller Challenge
http://challenge.bfi.org
Annual, jury-awarded $100,000 prize for "a bold, visionary, tangible initiative that is focused on a well-defined need of critical importance.... Solution must be part of an integrated strategy dealing with key social, economic, environmental, and cultural issues." Inspiring Web site.

Core77
http://www.core77.com
Design magazine and resource on schools, design firms, and jobs.

Coroflot
http://www.coroflot.com
Design jobs and portfolios.

Creative Generalist
http://creativegeneralist.blogspot.com
A blog for curious divergent thinkers who appreciate new ideas from a wide mix of sources.

David Byrne
http://www.davidbyrne.com
News and art from the musician and artist David Byrne.

Design Corner
http://designcorner.blinkr.net
Online archive for design news and posts. Lots of cool product design photos.

Design Observer

http://designobserver.com

The preeminent blog about all things design related, edited by Michael Bierut, William Drenttel, and Jessica Helfand, three of the best in the business.

Design That Matters

http://designthatmatters.org

Headed by Timothy Prestero, DtM is a nonprofit based in Cambridge, Massachusetts, which has built a collaborative design process through which hundreds of volunteers in academia and industry donate their skills and expertise to the creation of breakthrough products for global communities in need.

Design Thinking

http://designthinking.ideo.com

Thoughts from IDEO CEO Tim Brown.

Design 21

http://www.design21sdn.com

Exploring ways that design can positively impact our communities—ways that are thoughtful, informed, creative, and responsible. Also, see their competition page.

Designers Accord

http://www.designersaccord.org

A global coalition of designers, educators, researchers, engineers, and corporate leaders, working together to create positive environmental and social impact.

Designism Connects

http://www.designismconnects.org

Connecting creative people to causes for social change.

Doors of Perception

http://www.doorsofperception.com

An international conference and knowledge network founded by John Thackara that sets new agendas for design—in particular, for information and communication technologies (ICTs).

Dori Tunstall

http://dori3.typepad.com

Explorations in design, decision-making, anthropology, and policy.

Full Belly Project

http://www.fullbellyproject.org

Founded by Jock Brandis, designer of the universal nut sheller, the Full Belly Project designs and distributes income-generating agricultural devices to improve life in developing countries.

Gadget Nation

http://www.gadgetnation.net

Book site from author Steve Greenberg and Web store for gadgets.

Gizmodo

http://gizmodo.com

A popular gadget blog.

Gizmag

http://www.gizmag.com

Both a printed magazine and Web site about the latest technology and gadgets.

A Glimmer of Hope

http://www.aglimmerofhope.org

Austin, Texas–based foundation tackling the issue of global poverty, particularly in Ethiopia, employing business efficiencies.

Gordon Murray Design
http://www.gordonmurraydesign.com
Official site of the designers of the T.25
urban car.

In Good We Trust
http://www.ingoodwetrust2010.com
Mau's Web site for the 2010 Denver Biennial.

The Index:Award
http://www.designtoimprovelife.dk
Biennial presentation of the world's largest
award for design, a € 500,000 purse divided
evenly among winners in five categories: body,
home, work, play, and community. A global
competition sponsored by the country of
Denmark.

Innocentive
http://www.innocentive.com
Connecting companies, academic institutions,
the public sector, and nonprofit organizations
with a global network of the world's brightest
minds. InnoCentive blog offering perspectives
on innovation can be found at http://blog.
innocentive.com.

Make Magazine
http://makezine.com
Blog and magazine dedicated to DIY
technology projects.

Mau's Incomplete Manifesto for Growth
http://www.brucemaudesign.com/incomplete_
manifesto.html
Written in 1998, Mau's Incomplete Manifesto
for Growth is an articulation of statements
exemplifying Bruce Mau's beliefs, strategies,
and motivations.

Metacool
http://metacool.typepad.com
Thoughts on the art and science of bringing
cool stuff to life, by Diego Rodriguez.

Mixergy
http://blog.mixergy.com
Aimed at creatives and Web visionaries,
Mixergy hosts networking mixers, educational
forums, and online interviews.

19.20.21.
http://192021.org
"19 cities in the world, 20 million people, 21st
century." Five creative partners, including
Richard Saul Wurman, have launched a global
multimedia initiative to understand
population's effect on urban and business
planning, and consumers. Great Web site
graphics (no surprise).

Nussbaum on Design
http://www.businessweek.com/innovate/
NussbaumOnDesign
Short but highly informative blog postings on
the latest design and innovation news from
Bruce Nussbaum, influential journalist at
BusinessWeek.

Procrastineering
http://procrastineering.blogspot.com
Project blog interested in creating enabling
techniques that can significantly increase the
accessibility of technology, from Johnny Chung
Lee, a researcher in the applied sciences group
at Microsoft.

Taproot Foundation
http://taprootfoundation.org/blog

Blog for those interested in integrating the pro bono ethic into their careers—giving time and talent to strengthen nonprofit organizations.

Project M
http://www.projectmlab.com
Intensive summer program designed to inspire young graphic designers, writers, photographers, and other creative people so that their work can have a positive and significant impact on the world.

Springwise
http://www.springwise.com
Network of eight thousand spotters scanning the globe for smart new business ideas, delivering inspiration to entrepreneurial minds.

Unbeige
http://www.mediabistro.com/unbeige
A wide-ranging blog about design.

Wired to Care
http://www.wiredtocare.com
Dev Patnaik's tales of companies, political campaigns, sports teams, governments, and institutions of every kind that use empathy to connect with customers and employees.

X Prize Foundation
http://www.xprize.org
The X PRIZE Foundation is generating innovative breakthroughs in new and exciting areas and plans to launch ten new prizes over the next five years with a combined purse of approximately $100 million. Prize areas are Energy and Environment, Exploration, Education and Global Development, and Life Sciences.

COPYRIGHT AND PATENT INFO

BOOKS

From Edison to iPod: Protect Your Ideas and Make Money, Frederick W. Mostert and Lawrence E. Apolzon, Dorling Kindersley, 2007

ORGANIZATIONS

U.S. PATENT AND TRADEMARK OFFICE (USPTO)

General Information
http://www.uspto.gov
800-786-9199 or 571-272-1000

Inventor Assistance Center
800-786-9199 or 571-272-1000

Trademark Assistance Center
Madison East, Concourse Level
600 Dulany Street, Alexandria, VA 22314
800-786-9199 or 571-272-9250

THE AMERICAN BAR ASSOCIATION SECTION OF INTELLECTUAL PROPERTY LAW

American Bar Association
321 North Clark Street
Chicago, IL 60654-7598
800-285-2221
Free help with intellectual property: http://www.abanet.org/intelprop/probono.html
List of intellectual property sites:
http://www.abanet.org/intelprop/sites.html

notes

THE BRIEFING

Page 5: "Design for design's sake." Phil Patton quotes from his book *Michael Graves Designs: The Art of the Everyday Object* (Melcher Media, 2004), page 42. (Speaking of the Juicy Salif, Patton wrote, "Functionally, the juicer was a flop." Design critic Donald Norman has also observed that the Salif was more of an aesthetic triumph than a functional one.)

Page 5: "The arms race of designer toilet brushes." Said by *A Whole New Mind* author Dan Pink in "Why the Right Brain Rules," an *Advertising Age* interview with Jonah Bloom, 7/11/06.

Page 6: "Everything that I designed is absolutely unnecessary." Philippe Starck quoted in an interview with the German newspaper *Die Zeit*, 3/27/08.

Page 6: "Rumblings of a backlash against nonessential 'stuff' proliferated by designers." The backlash was discussed in a posting "Beware the Backlash" by Kevin McCullagh on the core77 blog, 1/07.

Page 7: "Nearly two hundred thousand new [products] introduced each year." Source is Mintel Market Research's 2007 Global New Products Database, www.mintel.com.

Page 7: The "innovation gap" is a term used by Patrick Whitney of IIT's Institute of Design, as explained in Chapter 4.

Page 8: "Businesspeople must begin to think like designers." Roger Martin said this in my interview with him, though this quote from him first appeared in *Fast Company*, 10/06.

Page 8: Heuristic bias concept explained to me by John Bielenberg of Project M, as well as in *How* magazine's article "Project M," by Alissa Walker, 2/06.

Page 13: "Several of [the principles] derive from Mau's original manifesto." For the record, the principles *Ask stupid questions*, *Go deep*, *Work the metaphor*, and *Begin anywhere* appeared in Mau's manifesto. *Jump fences* is a variation on Mau's *Avoid fields. Jump fences. Design what you do* is also Mau's, but did not appear in the manifesto. *Make hope visible* comes from Brian Collins. *Design for emergence* is from Greg Van Alstyne. *Face consequences* and *Embrace constraints* are not attributed to any single designer.

Page 15: The designer who advised "steering into a skid" is Marty Cooke, and the quote is used in full and explained later in the book, in Chapter 6, pages 177–78.

CHAPTER 1: ASK STUPID QUESTIONS

Page 23: Definitions of "framing" culled from my discussions with Steve McCallion of Ziba Design, Tim Brown of IDEO, and George Kembel of Stanford University's d.school.

Page 25: "Modo has been known to use cue cards . . . to continually remind employees . . . to ask 'why' at every stage of conducting business." From *Inc.* magazine profile of Modo, 10/02.

Page 25: "IDEO has established a methodology practice known as the Five Whys." From my interview with IDEO executive Jane Fulton Suri and from IDEO's "method cards."

Page 27: The OXO story is from my interviews with the founders of Smart Design. A version of this story also appeared in the book *Do You Matter? How Great Design Will Make People Love Your Company,* by Robert Brunner and Stewart Emery with Russ Hall (Pearson Education/FT Press, 2008).

Page 29: "Design is . . ." chart IA Collaborative. Sources of definitions: Bierut, Lois, Mau, Collins, Vignelli, Isley, Scher, Behar, Duffy, Mok, and Glaser definitions all were told to me by these sources during my interviews for this book. One of Glaser's definitions is adapted from the social scientist Herbert Simon, who wrote in his book *Sciences of the Artificial* that design is "the transformation of existing conditions into preferred ones." Sagmeister definition originally appeared in the book *How to Think Like a Great Graphic Designer,* by Debbie Millman (Allworth Press, 2007), reused here with Sagmeister's approval. Jobs definition appeared in *Fortune* magazine, 1/24/00, and was originally stated as: "Design is the fundamental soul of a man-made creation that ends up expressing itself in successive outer layers of the product or service." Papanek definition appeared in his book *Design for the Real World* (Academy Chicago Publishers, 1984). Lupton definition from her book *D.I.Y: Design It Yourself* (Princeton Architectural Press, 2006). Noyes definition from the book *Eliot Noyes,* by Gordon Bruce (Phaidon Press, 2007). Eames definition has been quoted many times in many places, including by Brian Collins at the World Economic Forum in Davos, January 2008.

Page 30: "Design has to work. Art does not." Quote from Donald Judd most recently featured at Cooper-Hewitt's National Design Triennial, 2006. "Art has to move you and design does not, unless it's a good design for a bus." David Hockney quote cited by Michael Bierut in *Seventy-nine Short Essays on Design* (Princeton Architectural Press, 2007), taken from *Design Is Not Art* catalog.

Page 30: "Art design" (or "design art") was written about by design critic Alice Rawsthorne in the *International Herald Tribune,* 1/22/07.

Pages 32–34: Donald Norman "wandered into design after first wading into a disaster." From my interview with Norman about how his experience investigating Three Mile Island sparked his interest in design. Also discussed in Norman's book, *The Design of Everyday Things* (Basic Books, 1988, 2002).

Page 33: Donald Norman's three levels of design were discussed in my interview with him and are explained in more detail in his book *Emotional Design: Why We Love (or Hate) Everyday Things* (Basic Books, 2005).

Page 34: "The ideal response to a designed product might go something like this." The three responses that follow are my wordings, not Norman's.

Page 34: "Beautiful design . . . 'makes the medial prefrontal cortex light up, just like sex.'" The former Starbucks (now Google) designer Robert Wong told me this in an interview and it is also featured in his 2006 booklet, "Eight Unwritten Principles of Design."

Page 34: "Some of our visceral responses to design seem to be hardwired." From my interview with Moshe Bar. Bar also wrote a good article on this, "To Get Inside Their Minds, Learn How Their Minds Work," in *Advertising Age*, 11/26/07.

Page 35: "If something is beautiful, it may be easier to use." Norman elaborates on this point in *Emotional Design*, citing a study by Japanese researchers Masaaki Kurosu and Kaori Kashimura, page 17.

Page 37: "Can we please have our name back?" Mike Dempsey is quoted saying this in the *Observer* article "Design: Politics of the drawing board," by Geraldine Bedell, 11/27/05.

Page 37: "Design is presented as a kind of transformative cure-all." Michael Bierut as quoted in *Creativity* magazine, 4/05.

Page 38: Milton Glaser told me this story in an interview with him. He told a similar version of the story in Debbie Millman's book *How to Think Like a Great Graphic Designer* (Allworth Press, 2007).

Page 38: "I didn't see why my bum had to hurt." Thomas Heatherwick said this in an interview with *Time* magazine for its Style & Design 100 special issue, 2007.

Page 39: "Not everything is about design, but design is about everything." From my interview with Michael Bierut in *One* magazine, Volume 11.3, Winter 2008.

Page 40: "I don't want to die here—it's too ugly." Michael Graves quote cited in *AARP Bulletin* article by Louise Sloan, 9/06.

CHAPTER 2: JUMP FENCES

Page 46: Abductive reasoning is discussed in the *BusinessWeek* article "Creativity That Goes Deep," by Roger Martin, 8/3/05.

Pages 46–47: Kamen's artificial hand was demonstrated for me (including a viselike handshake) during my visit to DEKA Research.

Page 47: Roger Martin's explanation of how and why designers think they can solve any problem, from my interview with him and also from the *BusinessWeek* article "The Positive Spiral: Six Keys to Success," by Roger Martin, 2/28/07.

Page 48: John Thackara's definition of "smart recombinations" comes from his book *In the Bubble: Designing in a Complex World* (MIT Press, 2005), page 217.

Pages 48–49: Van Phillips's description of how he connected unrelated ideas to come up with the Cheetah foot comes from my interview with him and also from a *New York Times* article, "A Personal Call to a Prosthetic Invention," by Carol Pogash, 7/2/08.

Pages 49–50: "Researchers say the brain's right hemisphere is fertile ground for such far-ranging, hopscotching activity." From Jonah Lehrer's excellent *New Yorker* piece "The Eureka Hunt," 7/28/08.

Page 53: "There is something of a schism in the design world." Described in the *BusinessWeek* article "Masters of Collaboration," by Steve Hamm and Helen Waters, 1/16/08.

Page 54: Cynthia Barton Rabe's discussion of "Zero-Gravity Thinkers" is from her book

The Innovation Killer: How What We Know Limits What We Can Imagine (AMACOM, 2006).

Page 54: "The curse of knowledge." This phrase has been used by, among others, Chip and Dan Heath in their book *Made to Stick: Why Some Ideas Survive and Others Die* (Random House, 2007).

Pages 55–61: The description of the design process of the XO laptop comes from my interviews with Yves Behar and Bret Recor of fuseproject, Nicholas Negroponte of MIT, and Kevin Young of Continuum. An excellent article on the subject appeared in *Wired* magazine, "The Laptop Crusade," by Douglas McGray, 8/06.

Page 56: Negroponte's equations (cheap laptop vs. inexpensive laptop) were shared with me in an e-mail interview conducted with him.

Page 58: "The effort was publicly dismissed . . . by Intel and Microsoft." Page 67: "Intel . . . began to try to lure away customers." The behavior by XO laptop competitors was described in the *Wall Street Journal* ("A Little Laptop with Big Ambitions," 11/24/07) and the *New York Times* ("At Davos, the Squabble Resumes on How to Wire the Third World," 1/29/07; "Intel Quits Effort to Get Computers to Children" 1/5/08).

Page 58: "Dinosaur baby." Used in a quote by IDEO's Paul Bennett in *Advertising Age* magazine, 9/11/06.

Page 62: "Designers are needlessly constrained by the myth that everything they do must be a unique and creative act." From John Thackara's book *In the Bubble*.

Page 62: "Michael Bierut's brain 'is a compendium.'" Paula Scher's description of her Pentagram partner's brain can be found in the profile of Bierut on the AIGA.org Web site (in the Design Archives/Medalists section).

Pages 63–64: Stefan Sagmeister's quotes are from my interview with him; some appeared originally in my article about him in *One* magazine, Volume 11.4, Spring 2008.

Page 63: "De Bono found that the brain has a natural propensity for repetition." Sagmeister is citing ideas from Edward de Bono's books, including *Lateral Thinking: Creativity Step by Step* (Perennial Library, 1973) and *The Thinking Course* (Facts on File, 1994 revision).

Page 64: A description of how Tibor Kalman intentionally designed things "the wrong way" appeared in the *Step Inside Design* article "Wrongness in the Walls: M&Co. Remembered," by Tiffany Meyers, May/June 2006.

Page 65: Recent studies on insights conducted by Northwestern University professor Mark Jung-Beeman. From the *New Yorker* article "The Eureka Hunt," 7/28/08.

Page 67: "Intel . . . began to try to lure away customers." *See page 58 note.*

Page 67: "The End of the One Laptop Per Child Experiment: When Innovation Fails." This post appeared on Bruce Nussbaum's blog Nussbaum On Design, 5/16/08. A subsequent *BusinessWeek* cover story ("One Laptop Meets Big Business," 6/5/08) followed in which critic William Easterly charged laptop makers with cultural imperialism and being "arrogant."

Page 67: "The normally cool, laid-back Yves Behar looked like he might blow a gasket." My observation at ICFF in New York on May 19, 2008.

CHAPTER 3: MAKE HOPE VISIBLE

Page 72: "Everybody has ideas about how to fix things." From an interview I did with Clement Mok that originally appeared in *One* magazine, Volume 11.3, Winter 2008.

Page 73: "Sketching is 'the archetypal activity of designers.'" From Bill Buxton's *Sketching User Experiences: Getting the Design Right and the Right Design* (Morgan Kaufmann, 2007), page 111.

Page 74: Milton Glaser's description of exploratory sketching taken from my interview with him, also described in much more detail in his book *Drawing Is Thinking* (Overlook, 2008).

Page 75: "We acquire far more information through vision than all other senses combined." From my interviews with Colin Ware of the Data Visualization Lab. More can be found in Ware's books *Information Visualization* (Morgan Kaufmann, 2004) and *Visual Thinking for Design* (Morgan Kaufmann, 2008).

Page 76: Daniel J. Simons's video, in which the "gorilla in our midst" is hard to detect, can be viewed at this link: http://viscog.beckman.illinois.edu/flashmovie/15.php.

Page 76: "There will be a *billionfold* increase in the amount of visual stimuli." From Bruce Mau's book *Life Style* (Phaidon Press, 2000).

Page 77: Michael Bierut's observation that Vignelli's subway map "delivered the necessary information at the point of decision" appeared in *Seventy-nine Short Essays on Design*, page 136.

Page 77: "The rise of smartphones has heightened the need to compress and distill visual data." The idea of stripping down phones to basic functions was discussed in "On a Small Screen, Just the Salient Stuff," by John Markoff in the *New York Times*, 7/13/08.

Page 78: John Maeda's description of "thoughtful reduction" in design appears in his book *The Laws of Simplicity* (MIT Press, 2006), page 1.

Page 78: "The mind loves puzzles." From my interview with Milton Glaser, also covered in a Forbes.com interview with Glaser by Lacey Rose, 9/28/05.

Page 79: Richard Saul Wurman's "Are mammograms important?" chart originally appeared in his book *Understanding Healthcare* (TOP, 2004) and is reprinted with his permission.

Pages 79–80: The analysis of the use of design in Al Gore's film *An Inconvenient Truth* was shared with me separately by both John Maeda and Brian Collins.

Pages 81–82: The creation of the WE logo was described to me by Brian Collins. Steven Heller also wrote about the making of the logo in the *New York Times* article "Al Gore's New Logo," 4/6/08.

Pages 83–86: Deborah Adler's story comes from my interviews with her. The story has also been told elsewhere, including *New York* magazine's article "The Perfect Prescription," by Sarah Bernard, 4/11/05.

Page 87: Tufte has talked about the *Challenger* charts in his book *Visual Explanations* (Graphics Press, 1997).

Page 93: One critic compared Massive Change to "a school project." The criticism of Massive

Change was reported in "How Now, Bruce Mau?" by Pamela Young, *Applied Arts* magazine, 3/06.

Page 93: Richard Lacayo's review of Mau's Massive Change show in Vancouver appeared in *Time* magazine, 10/18/04.

Page 94: "A film production team affiliated with Leonardo DiCaprio." This was for the 2007 film *The 11th Hour*, which featured DiCaprio as the narrator and Mau as one of the commentators.

CHAPTER 4: GO DEEP

Page 99: "Nearly two hundred thousand products." This is from aforementioned Mintel Global New Products Database.

Pages 100–101: "If I'd simply asked customers what they wanted, they would have said 'a faster horse.'" This is an oft-cited quote by Henry Ford, reflecting back on the manufacture of his first car.

Pages 102–103: "The Three Gears of Design" is from my interview with Heather Fraser, and explored further in the *Globe and Mail's* article "Getting From 'Design Thinking' to 'Design Doing,'" by Heather Fraser, 5/9/06.

Page 104: The observation that *innovation* is a more "masculine" term, while *design is "associated with* drapes" appeared in Bruce Nussbaum's blog Nussbaum On Design, 3/18/07.

Page 104: "Innovation is the new black." Michael Bierut's *Seventy-nine Essays*, page 217.

Pages 106–107: "Design-driven companies outperformed the Standard & Poor's 500 by a ten-to-one margin." From a three-year study titled "Customer Experience (cX) Design for Services," released 12/07 by Peer Insight, a Washington, D.C.–based research firm.

Page 107: "Design-driven British companies outperformed competitors by 200 percent." From the UK Design Council's groundbreaking 2005 Design Index report, which analyzed 150 companies' results between 1994 and 2003.

Pages 108–109: "The 'missing toes' incidents." From my interviews with Jane Fulton Suri.

Pages 110–111: Early days of IDEO from interviews with David Kelley, Jane Fulton Suri, and others at IDEO.

Page 111: "Your meat loaf tastes like sawdust." From *The Art of Innovation: Lessons in Creativity from IDEO, America's Leading Design Firm* by Tom Kelley with Jonathan Littman (Currency/Doubleday, 2001), page 27.

Page 112: "Unfocus groups." Definition from IDEO method cards and a *New York Times* article, "Going Off the Beaten Path for New Design Ideas," by Lisa Chamberlain, 3/12/06.

Page 112: "The firm's designers spent time flat on their backs in hospital beds." IDEO's Paul Bennett in his essay "Make Consumers Part of the Design Process by Tuning In" in *Creativity* magazine, 3/6/06.

Page 112: "A blinding glimpse of the bleeding obvious." IDEO's Paul Bennett, *ibid*.

Page 114: "A cache of buried Stone Age hand tools was unearthed recently." From the *New York*

Times article "Colorado Backyard Yields Cache of Stone Age Tools," by Kirk Johnson, 2/26/09.

Page 115: "You couldn't do better than watching a transvestite." Discussed in the *Times* (London) article "You Know When It Feels Like Somebody's Watching You . . . ," by Rhys Blakely, 5/14/07.

Pages 115–116: "Rubbermaid was developing a new line of walkers for senior citizens." Rubbermaid's use of observation in designing new walkers is covered in *Innovation: Breakthrough Thinking at 3M, DuPont, GE, Pfizer, and Rubbermaid,* edited by Rosabeth Moss Kanter, John J. Kao, and Fred Wiersema (Collins, 1997).

Page 117: "The 'butt brush' effect." Described in Paco Underhill's book *Why We Buy: The Science of Shopping* (Simon & Schuster, 1999).

Page 117: "An authorized stalker." Jan Chipchase's self-description in the *Times* (London) article "You Know When It Feels Like Somebody's Watching You . . . ," by Rhys Blakely, 5/14/07.

Page 118: "And ethnography can, indeed, occasionally go to those titillating extremes." Story about Unilever research in "Making Market Research Cool," by Jack Neff in *Advertising Age,* 4/28/08.

Page 123: "You have folks marketing tomato sauce who dine at fine Italian restaurants . . . " From "Companies from Mars, Customers from Venus," an interview with Dev Patnaik by Elizabeth Olson in the *New York Times,* 1/24/09.

CHAPTER 5: WORK THE METAPHOR

Pages 126–130: The Umpqua story is from my interviews with CEO Ray Davis, Umpqua creative director Lani Hayward, Steve McCallion of Ziba Design, and Jim Haven of Creature design firm. Also, from the *Economist* article "Retail Banking: Branching Out," 6/16/07.

Page 130: "'Experience design,' a concept that has been circulating in the business world since the 1990s." The 1999 book *The Experience Economy: Work Is Theater & Every Business a Stage,* by B. Joseph Pine and James H. Gilmore (Harvard Business School Press), has been a big driver of the concept.

Page 139: Mau's description of the Indigo store design originally appeared in his book *Life Style* (Phaidon, 2000).

Page 140: "The transformational metaphor [at Starbucks] was an Italian espresso bar." Starbucks founder Howard Schultz describes his espresso bar epiphany at length in his book *Pour Your Heart Into It* (Hyperion, 1997).

Page 140: "As more of our basic needs are met, we increasingly expect sophisticated experiences." From Tim Brown's article "Design Thinking," in the *Harvard Business Review,* June 2008.

Page 141: "People want an uplifting experience so they can persuade themselves they're not doing a grubby self-interested transaction." From David Brooks's column titled "Questions for Dr. Retail" in the *New York Times,* 2/8/08.

Page 143: Mihály Csíkszentmihályi's seminal book on immersion is *Flow* (HarperCollins, 1990).

Page 143: "A 'stop time' effect." From "It's Time to Understand Consumers," by John Rosen and AnnaMaria Turano in *Advertising Age*, 2/4/08.

Page 144: On experience designers' need to follow a "North Star." From Brandon Schauer, an experience design director at Adaptive Path, and his Google Tech Talk, 5/8/08.

Pages 145–146: "Google's designed experience." From my interview with Google's Irene Au.

Page 146: "What's required is 'a quantum evolution—beyond technology and features.'" Peter Merholz, president of Adaptive Path, in his essay "Experience *is* the product—and the only thing users care about," on the Core77 blog, 6/07.

Page 148: "Martin believed that a big, splashy TV ad campaign wasn't really necessary." From my interview with Kerri Martin in *One* magazine, Volume 6.2, Fall 2002.

Page 149: Gert Volker Hildebrand on the Mini design, as quoted in *Eureka* magazine by Tom Shelley, 8/18/08.

Page 149: "Mini spends less than 1 percent on design." *Ibid.*

Page 151: "If you actually set out to design a dream product." From my interview with Alex Bogusky in 2006 for the book *Hoopla* (Powerhouse, 2007).

Page 152: "Smoking among middle and high school students in Florida had declined an average of 38 percent." From the 2002 Florida Youth Tobacco Survey, Florida Department of Health.

CHAPTER 6: DESIGN WHAT YOU DO

Page 155: The Coke anecdote took place at the Promax BDA (design and broadcast) conference in New York on 6/18/08.

Page 158: "Unilever had to answer for why it was celebrating . . ." This was covered in many articles, including, "Advocacy Group Blasts Unilever's 'Hypocrisy'" by Kamay High, *Adweek*, 10/11/07.

Page 159: "Mau . . . had grappled with the concept of 'incorporation.'" "Incorporations" was the theme of the book *Zone 6* (Urzone, 1992).

Page 161: "The Coke brand had become like cultural wallpaper." From my interview with David Turner in *One* magazine, Volume 12.2, Fall 2008.

Page 162: "Only in the past few years has Coke tried to rediscover its design roots." From the *BusinessWeek* article "Coke Works Through Identity Crisis," by Jessie Scanlon, 8/26/08.

Page 163: "The sin of the hidden trade-off." From the *Fast Company* article "Another Inconvenient Truth," by David Roberts, 3/08.

Page 164: "UPS, which has redesigned its truck delivery systems." From the *Boston Globe* article "No Left Turns Is Actually Right On," by Peter DeMarco, 4/20/2008.

Pages 166–171: The Pedigree story is from my interview with Lee Clow. Portions of the interview also appeared in *One* magazine, Volume 12.2, Fall 2008.

Pages 171–174: The description of how Procter & Gamble introduced design thinking to the company was the subject of a series of interviews at the IIT Institute of Design's 2008 Strategy

Conference, held 5/23/08 in Chicago. Lafley was interviewed by Roger Martin, and also interviewed briefly by me afterwards; most of Kotchka's remarks were given to the group, some to me when questioned afterwards.

Page 175: "A rabbi inside the corporation." Milton Glaser quote cited by design writer Ralph Caplan in the introduction to the School of Visual Arts "master series" tribute to Steven Heller in July 2007.

Page 177: "Culture of courage." From Heather Fraser's article "Turning Design Thinking into Design Doing" in *Rotman Magazine*, Spring/Summer 2006.

Page 177: "Culture of optimism." From an essay titled "Creating an Entrepreneurial Culture of Optimism," by IDEO's Brendan Boyle for ABC News, 1/14/08.

Page 177: "An ecosystem of ideas where we just build stuff and see if it works." Au's description of the culture of Google is from my interview with her, though the "ecosystem" quote appeared in a profile of her by Chuck Salter on FastCompany.com, 2/14/08.

Page 177: "Steering into a skid—it's counterintuitive." Quote by Marty Cooke of the ad/design firm SS+K cited in "How Creativity Can Carry Your Business Through a Recession," by Teressa Iezzi in *Advertising Age,* 9/22/08.

Pages 177–178: "If you tighten up too much, you eliminate future innovation." Google's Eric Schmidt interviewed in "Google at 10: Searching Its Own Soul," by Miguel Helft in the *New York Times*, 11/8/08.

Page 179: "70 percent of consumers feel business bears responsibility for driving positive social change." From a study by Euro RSCG, published in the article "Consumers Are Watching You," *Advertising Age*, 4/7/08.

Page 179: "In a separate study, by Condé Nast, respondents said a company's overall behavior could and should be judged by four essential characteristics." Study results cited in the article "'Good' Matters to Consumers," *Advertising Age*, 12/10/07.

Page 180: "The selfish interests and the ethical interests are coming together." Quote from Bill Buxton from "Design Takes Over, a Fly Swatter at a Time," by Sarah Milroy in the *Globe and Mail*, 4/23/07.

CHAPTER 7: FACE CONSEQUENCES

Pages 183–184: Sinclair's presentation was part of the *Metropolis* magazine conference "Design Entrepreneurs: Make Good and Prosper," held during the ICCF show in New York on 5/19/08.

Page 185: "The population segment that design activists have dubbed 'the other 90 percent.'" This is a reference to the exhibition Design for the Other 90%, which originally ran in New York from May to September 2007 at the Cooper-Hewitt National Design Museum and continues to exhibit at museums around the world. An accompanying 114-page catalog was also published.

Pages 187–190: Most of Cameron Sinclair's story comes from my interviews with him, but some descriptions of AFH's early contests came from his book *Design Like You Give a Damn*, edited with Architecture for Humanity (Metropolis Books, 2006).

Pages 193–196: The story of rebuilding homes in Biloxi comes from my interviews with Cameron Sinclair and architect Marlon Blackwell. The Biloxi rebuilding effort was also covered in the *New York Times* article "Design Steps Up in Disaster's Wake," by Allison Arieff, 8/2/07.

Page 196: The RISD kitchen initiative, including the study showing the three hundred steps needed to make spaghetti, was described to me by former RISD president Roger Mandle. The kitchen, no longer in use, has been displayed at the Cooper-Hewitt Museum.

Pages 197–198: My description of the RISD students' efforts to design a dining hall in Tanzania comes from attending the final presentation of their project to professor Liliane Wong at RISD on 5/22/08.

Page 200: Buckminster Fuller's "egocide" story was included in the exhibit Buckminster Fuller: Starting with the Universe, Whitney Museum of American Art, 6/26–9/21/08.

Page 201: "We've realized our dreams at the expense of Lake Michigan." Charles Eames's quote is from *New York* magazine's article "Architecture: Designer Marriage," by Karrie Jacobs, 9/6/99.

Page 201: "All those best-laid plans went seriously astray." Again, from Richard Lacayo's review of Massive Change, *Time* magazine, 10/18/04.

Page 203: "The iPod moment." From Jakob Trollbäck's essay "Surface vs. Purpose," *Creativity* magazine, 4/06.

Page 203: "We're filling the world with stuff." From John Thackara's book *In the Bubble*.

Pages 203–204: The Designism 3.0 event I attended was held at the Art Director's Club in New York on 10/2/08.

Pages 204–205: My information about the Aquaduct bike was obtained partly at the Designism event, but also via separate e-mail interviews conducted with one of the designers, Adam Mack, for an article appearing in *One* magazine, Volume 11.4, Spring 2008.

Page 205: "We need a hundred thousand Dean Kamens." Thomas Friedman said this at the Aspen Ideas Festival, Aspen, Colorado, on 7/3/08.

Page 208: The Thackara-Sinclair exchange took place on the Design Observer blog in July 2008.

Page 209: Thomas Fisher's "fracture critical" theory was explained to me in an interview with Fisher and is explored in greater depth in Fisher's book *Architectural Design and Ethics: Tools for Survival* (Architectural Press, 2008).

CHAPTER 8: EMBRACE CONSTRAINTS

Page 215: "This might be the time when designers can really do their job." Paola Antonelli's quotes are from the *New York Times* article "Design Loves a Depression," by Michael Cannell (1/4/09), as well as from a follow-up interview with me in which she elaborated on this theme.

Page 217: "Satisficing." Definition of satisficing is partially from the book *Universal Principles of*

Design: 100 Ways to Enhance Usability, Influence Perception, Increase Appeal, Make Better Design Decisions, and Teach Through Design, by William Lidwell, Kritina Holden, and Jill Butler (Rockport Publishers, 2003).

Pages 217–218: "Ephemeralization." Described at the 2008 Buckminster Fuller exhibition Starting with the Universe at the Whitney Museum. Also discussed in the exhibit's 258-page catalog, *Buckminster Fuller: Starting with the Universe,* by K. Michael Hays (Yale University Press, 2008), and many other books about Fuller.

Page 221: Design for the Other 90%. This exhibition ran in New York from May to September 2007 at the Cooper-Hewitt National Design Museum and continues to exhibit at museums around the world. An accompanying 114-page catalog was also published.

Page 221: "The so-called bottom of the pyramid." Refers to developing countries, as termed in the book *The Fortune at the Bottom of the Pyramid: Eradicating Poverty Through Profits,* by C. K. Prahalad (Wharton School Publishing, 2006).

Page 221: "A billion customers in the world are waiting for a $2 pair of eyeglasses." This quote from Dr. Paul Polak appeared in the *New York Times* article "Design That Solves Problems for the World's Poor," by Donald G. McNeil Jr., 5/29/07.

Pages 222–223: Martin Fisher was interviewed by *Glimmer* researcher Tiffany Meyers. Fisher's "price of a chicken" quote originally appeared in the Design for the Other 90% exhibition catalog, page 35.

Page 222: "KickStart pumps have helped more than four hundred thousand people escape from poverty." KickStart's impact report is posted on its Web site, KickStart.org.

Page 223: "Designers in emerging markets are often forced by severe price constraints to try to improve efficiency by 90 percent." From *The Fortune at the Bottom of the Pyramid* and also from Tiffany Meyers's interview with C. K. Prahalad for *Glimmer*.

Pages 223–224: "Strength from weakness." Shigeru Ban's story of building refugee housing in Rwanda from paper is taken from my interview with him.

Page 225: "The new design is forever." Mau unveiled this slogan at the Promax BDA conference in New York on 6/18/08.

Pages 225–226: William McDonough was interviewed by Tiffany Meyers for *Glimmer*. Also, the quote "Does it have reverse logistics?" is from *Newsweek*'s article "Buildings That Can Breathe," by Fareed Zakaria, 8/18/08. For more of McDonough's ideas on sustainable design, see *Cradle to Cradle: Remaking the Way We Make Things*, by William McDonough and Michael Braungart (North Point Press, 2002).

Pages 229–230: Janine Benyus was interviewed by Tiffany Meyers for *Glimmer*.

Page 235: "A direct link between diversity and innovation." Scott E. Page's thoughts about creativity, innovation, and diversity were reported in a conversation with Claudia Dreifus in the *New York Times* article "In Professor's Model, Diversity = Productivity," 1/8/08. Also Page's book *The Difference: How the Power of Diversity Creates Better Groups, Firms, Schools, and Societies* (Princeton University Press, 2007) explores the link between diversity and innovation.

Page 236: "But what did she design?" The controversy about Hilary Cottam receiving the 2005 Design Museum's Designer of the Year award was reported in several places, including the *Observer*'s article "Design: Politics of the drawing board," by Geraldine Bedell, 11/27/05.

CHAPTER 9: DESIGN FOR EMERGENCE

Page 245: "We have the freedom to create, in some small measure, the world in which we want to live." Fritz Haeg, quoted from his book *Edible Estates: Attack on the Front Lawn* (Metropolis Books, 2007), page 22, and reiterated in my interview with him.

Page 245: "The principle of designing for emergence." From Greg Van Alstyne and Robert K. Logan's research paper, "Designing for Emergence and Innovation," published in *Artifact* 1.2, 2007.

Page 249: "The psychologist Mihály Csíkszentmihályi studied what was most important to people in their homes." This was done by Csíkszentmihályi and Eugene Rochberg-Halton for their book *The Meaning of Things* (Cambridge University Press, 1981); also mentioned in Donald Norman's *Emotional Design* and discussed in my interview with Csíkszentmihályi.

Page 250: "Ted Baumgart, a film production designer in Los Angeles." Baumgart was originally interviewed by me for the book *Nextville: Amazing Places to Live the Rest of Your Life* (Springboard Press, 2008), pages 108–109.

Page 252: "Lawns must be 'maniacally groomed with mowers and trimmers.' " From Fritz Haeg's book *Edible Estates*, page 21.

Page 253: "The decline of card games, civic meetings, memberships in PTAs." From Robert D. Putnam's thesis in his book *Bowling Alone* (Simon & Schuster, 2001).

Page 253: "People have significantly fewer friends now than they did twenty years ago." From "Social Isolation in America: Changes in Core Discussion Networks Over Two Decades," a 2006 study issued by sociologists at Duke University and the University of Arizona.

Page 253: "Designed technology manages to connect and disconnect us at the same time." From my article "Me Me Media" in the August 2005 issue of *Reader's Digest*.

Page 255: "The fascinating new Action Center to End World Hunger in New York." Profiled in *One* magazine, Volume 12.2, Fall 2008.

Page 255: "Multiplayer games are the ultimate happiness engine." From Jane McGonigal's 2008 South by Southwest conference keynote speech.

Pages 256–258: Cottam's work on designing senior social circles from my interviews with her, and also covered in Alice Rawsthorne's *International Herald Tribune* article "A New Design Concept: Creating Social Solutions for Old Age," 10/26/08.

Page 257: "Informal 'Golden Girls' buddy networks." Creative co-housing solutions were explored in the book *Nextville* (Springboard Press, 2008), pages 67–71.

Page 259: "The older you get, the more you begin to think like a designer." Based on studies in the book *Progress in Brain Research: Essence of Memory* (Reed Elsevier, 6/08), as cited in the

New York Times article "Older Brain Really May Be a Wiser Brain," by Sara Reistad-Long, 5/20/08.

Page 259: "A series of T's linked together." The concept of connected T's is my own creation, based on discussions with Mau and Maeda.

Page 262: "Wurman and two partners started an architecture firm." From Richard Saul Wurman's book *Information Anxiety 2* (QUE, 2001), page 279.

Page 262: "If we are able to design our lives, wouldn't the best result—the best measure of success ultimately—be that every day is interesting?" From Richard Saul Wurman's book *Information Anxiety 2*, (QUE, 2001), page 286.

Page 263: "'How hearing works' images." Richard Saul Wurman's graphics originally appeared in his 2004 book *Understanding Healthcare* and are reprinted with his permission.

Pages 264–265: "People who are in a state of flow 'experience intense concentration.'" Mihály Csíkszentmihályi described flow in his interview with me and throughout his 1992 book, *Flow*. The Flow Chart originally appeared in his book on page 74 and is re-created here with his permission.

Page 264: "People tend to think of happiness as a goal, but it's more of a process." Martin E. P. Seligman's thinking about happiness appeared in the *Los Angeles Times* article "The Science of Happiness," by Marnell Jameson, 7/8/08, and in his 2003 book, *Authentic Happiness* (Nicholas Brealey Publishing).

Page 264: "It activates an area of the brain called the nucleus accumbens." From S. Ausim Azzizi's quote in the *New York Times* article "Hobbies Are Rich in Psychic Rewards," by Eilene Zimmerman, 12/2/07.

Page 265: "Being in that state of 'design flow' raises the levels of neurotransmitters." Observed by Dr. Gabriela Corá of the Florida Neuroscience Center, as quoted in the *New York Times* article "Hobbies Are Rich in Psychic Rewards," 12/2/07.

CHAPTER 10: BEGIN ANYWHERE

Page 270: "Designers as 'optimistic people who are trained to be courageous about the future.'" From Katie Rutter's 3/18/08 interview with Nathan Shedroff, as posted on the Adaptive Path blog.

Pages 271–275: The Stanford d.school process is derived from a series of interviews I did with George Kembel, one of the school's cofounders. The "Begin Anywhere" design process is my term for the process. The bucket chart is an original creation for *Glimmer*, based on Kembel's description of the process.

Pages 278–279: Todd Greene's HeadBlade story is from my interview with him. However, it should be noted that the HeadBlade was first reported on by Steve Greenberg in his excellent book *Gadget Nation: A Journey Through the Eccentric World of Invention* (Sterling, 2008).

Pages 279–280: My description of celebrity designers' activities is based on accounts in various news articles. Justin Timberlake's designing of a Memphis-area golf course was reported in the

New York Times article "Timberlake Sings the Praises of One of His Loves," by Bill Pennington, 8/4/08. Brad Pitt was quoted on his passion for architecture in *Time* magazine's "Verbatim" column, 6/16/08. Daniel Day-Lewis's furniture-making avocation was described in the *Wall Street Journal*'s "Sojourner in Other Men's Souls," 1/23/08.

Page 280: The description of David Byrne's design activities and the quotes from him are from an e-mail interview I conducted with Byrne. His bicycle rack designs were also reported on in the press, including the *New York Times* article "David Byrne, Cultural Omnivore, Raises Cycling Rack to an Art Form," by Ariel Kaminer, 8/8/08. Sketches of the bike rack and the chair designs are reprinted with Byrne's permission.

Pages 280–281: The story about Google's Patent Search and celebrity patents by the likes of Eddie Van Halen and Jamie Lee Curtis was first posted by Ryan Block on the blog IronicSans.com, 1/4/07, and then was widely circulated on other blogs.

Page 281: A "golden age for gadgets." This phrase appeared in the *New York Times* article "Fire, the Wheel, and, of Course, Mop Slippers," by Allen Salkin, 4/15/07.

Page 281: The MP3 Pez dispenser was covered in the article "Inventor Turns Pez Dispenser into MP3 Player," by Matt Hines, CNET News.com, 3/25/05.

Pages 281–282: Oliver Beckert's Aquariass was also featured in Steve Greenberg's book *Gadget Nation*.

Page 282: "To pull water from the air" refers to Aqua Sciences' system, which uses a blend of salts to collect water, as reported in the *Wall Street Journal*'s October 2007 article "2007 Technology Innovation Awards," by Michael Totty. "Get restaurants to stop overserving water" refers to the Tap Project, the brainchild of designer/adman David Droga, whose program encourages restaurant patrons to pay for glasses of tap water, with the money donated to clean water programs.

Page 283: "A Facebook-type network for seniors called Jive." Reported on by Bruce Nussbaum in "Ben Arent Creates a OLPG—One Laptop Per Grannie. Get Him Venture Funding," from a 6/23/08 posting on his blog, Nussbaum On Design.

Page 283: A bike "that doubles as an ambulance." Niki Dun's ambulance bike appeared in Bruce Mau's Massive Change exhibit. Her Web site is designfordevelopment.org.

Page 284: "Portland officials figured out that bike lanes weren't sufficient to protect bikers." From the *New York Times* article "Portland, Ore., Acts to Protect Cyclists," by William Yardley, 1/10/08.

Page 286: Blake Mycoskie's story comes from my own conversations with Mycoskie, but also has been reported in articles including *Time* magazine's "A Shoe That Fits So Many Souls," by Nadia Mustafa, 1/26/07.

Page 289: David Brooks's observations about focused attention appeared in his 12/16/08 column "Lost in the Crowd," in the *New York Times*.

Page 289: "One study, by Hewlett-Packard, found that constant interruptions actually sap intelligence." From a March 2005 survey of 1,100 British people, carried out by TNS Research and commissioned by HP.

acknowledgments

Glimmer could never have happened without the support of the designers I interviewed and spent time with. When a book author asks a successful and very busy person to give up his/her time, it is asking a lot. These people have no idea if the book is ever going to be finished, if they will like what's in it, or if it will get read by anyone. So it is an act of generosity to share one's time and ideas with an author, and the following designers were extremely generous: Brian Collins (who patiently answered all of my stupid questions, and when he couldn't make me understand with words, drew pictures for me); Yves Behar and others at fuseproject including Bret Recor and Melissa Guthrie; Dean Kamen; Cameron Sinclair; Hilary Cottam; Van Phillips; Janine Benyus; Marianne Cusato; Michael Bierut; Paula Scher; Stefan Sagmeister; IA Collaborative partners Kathleen Brandenburg and Dan Kraemer; Steve McCallion; Deborah Adler; Gordon Murray (and thank you to Will Bourne of *Fast Company* for connecting me with Gordon); Milton Glaser; Massimo Vignelli; Alexander Isley; George Lois; Joe Duffy; Clement Mok; Robert Wong; Shigeru Ban; ESI Design's Ed Schlossberg and Michelle Mullineaux; Gianfranco Zaccai and Kevin Young at Continuum; Dev Patnaik and Peter Mortensen at Jump Associates; William McDonough; Marlon Blackwell; Emily Pilloton; Martin Fisher; Blake Mycoskie; Steve Mykolyn; Jennifer Morla; and Richard Saul Wurman.

I want to say special thanks to two design firms. The welcoming and friendly folks at Smart Design (including Davin Stowell, Tom Dair, Dan Formosa, and Thomas Isaacson) opened their doors and let me see how design works, up close. And the team at IDEO—including Tim Brown, David Kelley, and the remarkable Jane Fulton Suri—were extremely generous with their time and wisdom (and Tim did this even though he was coming out with a design book around the same time as mine).

To write this book, I literally had to go to school because there was so much to be learned about design. I am extremely grateful to the Rhode Island School of Design for allowing me to spend time there, and I especially want to thank Liliane Wong and John Maeda. Patrick Whitney of IIT's Institute of Design kindly invited me to the school for one of its conferences, and also spent a good deal of time on the phone with me. Thank you to the Rotman School of Management, particularly Roger Martin and Heather Fraser, and also to Stanford University's d.school, especially George Kembel. Colin Ware at the University of New Hampshire's Data Visualization Research Lab, Thomas Fisher at the University of Minnesota, the Drucker School's Mihály Csíkszentmihályi (world's toughest name to spell), and John Bielenberg of the Project M workshops all contributed to my design education. I owe a particular debt of gratitude to Greg Van Alstyne of the Ontario College of Art & Design.

A number of business executives were helpful to me as I worked on the book. I want to particularly mention Jim Hackett of Steelcase, Irene Au of Google, Umpqua Bank's Ray Davis and Lani Hayward, Michelle Sohn at OXO, Chris Hacker at Johnson & Johnson, and Nate Pence at Method soap. Special thanks to Mark Strauss at FXFOWLE Architects, who shared insights and contacts.

Some great sources of inspiration and information: Paul Hughes and his thoughts/drawings on design thinking; the writings of Donald Norman and John Thackara; David Byrne's chair and bike rack designs; the ideas of Doblin's Larry Keeley, Adaptive Path, and Tom Monahan; the great journalistic coverage of design in *Creativity, Design Observer, Fast Company, Wired, Core 77*, and Bruce Nussbaum's blog; the good works of Maria Blair and the Rockefeller Foundation; Paola Antonelli's great design shows at MoMA; and the Cooper-Hewitt Museum's National Design Triennial.

Even more inspiring were the "everyday designers" featured in the book, including Mark Noonan, Todd Greene, Gauri Nanda, Eleanor Leinen, Jay Sorensen, and the wondrous Jock Brandis. Thanks to Steve Greenberg of *Gadget Nation* for introducing me to some of these people. And I want to salute Dwayne Spradlin and InnoCentive for encouraging and supporting the "basement Buckys" everywhere.

Special thanks to two great designers disguised as ad executives, Lee Clow and Alex Bogusky. It was Alex who, when asked who might be a good central character for this book, sent back an e-mail with just two words: Bruce Mau.

I am deeply indebted to Bruce for sharing so much time and so many great ideas. He opened up his studio and his home to me; his wife, Bisi, was extremely gracious and supportive. And I was greatly helped by a number of people at Bruce Mau Design, including Jim Shedden, Monica Bueno, Seth Goldenberg, and the amazing Randi (Fiat, that is).

This book started as an idea that was written up in one paragraph. I sent it to my literary agent, Sloan Harris at ICM and he encouraged me to pursue it. When I fleshed it out into a full proposal, Sloan read that on his way home from work and phoned from the train, full of enthusiasm. That, in turn, fueled my enthusiasm and the project was on its way.

Writers often say how lucky they are to have a good editor. Well, I was quadruply blessed. My Canadian editor, Craig Pyette, and my British editor, Nigel Wilcockson, both contributed great insights. My original Penguin Press editor, Vanessa Mobley, believed in this project from the outset and helped shape it. When Vanessa moved on, the book was taken over by Penguin Press founder Ann Godoff, a legend in the publishing business. I was nervous about working with an editor of her stature, but Ann put me at ease right away. She also gently pushed me, and the book, to a much higher level, for which I am so grateful.

Glimmer required a great deal of research, and Tiffany Meyers, an outstanding journalist with a strong knowledge of the design industry, helped by conducting some of the interviews and tracking down stories. The book also was very demanding from an editorial production standpoint, and Lindsay Whalen was a great asset in that area.

Finally, I want to thank all the people who provided encouragement and support along the journey of producing this book. That includes people who wrote wonderful letters and e-mails that boosted my spirits (Ann Collins and Kristyn Keene come to mind). Great support also came from The One Club for Art & Copy, in particular Mary Warlick and Yash Egami.

I want to thank my dear departed dad, who asked each time I visited, "So how's the book going?" It's done, Dad, and I wish I could raise a toast with you to celebrate.

Last and always first and foremost, thank you to my wife and creative partner, Laura E. Kelly, who had a hand in every aspect of this book, from the original idea to research, design elements, editing, and finally marketing and promotion. That's an awful lot for one person to take on, but not if that person is Laura. She is my inspiration, providing an endless source of "Glimmer moments" in my life.

Index

Page numbers in *italics* refer to illustrations.

abductive reasoning, 14, 45–50, 172
accessibility, 5–6, 44, 70–72, 200, 211–24, 249
accidents, 55
    prevention of, 108–10
action, 269–72
Action Center to End World Hunger, 255
adaptations, by customers, 115–16, 213, 220,
    231–32, 247
Adaptive Path, 144, 146
Adler, Deborah, 27, 83–86
advertising, 124, 158, 166, 167, *169,* 202, 244
    design as alternative to, 7, 12, 32, 81–82, 100,
        107, 114, 133, 147–50, 152–54, 157,
        162, 165
aesthetic-usability effect, 35, 258
aging, 8, 15, 24, 27, 113–14, 115–16, 124, 197,
    256–57, 258–59, 274, 282–83
AIDS, 188–89
AIGA, 86
aisles, 117, 140
alarm clocks, 276–77
Alliance for Climate Protection, 80–81
alternative energy, 6, 58, 59, 92, 194, 204–5, 212,
    213, 222, 245, 250–51, 285
amateur designers, 4, 16, 26, 44, 202–7,
    267–91
Amazon.com, 148
Antonelli, Paola, 215
Apple, 25, 72, 101, 105–6
    branding by, 15, 32, 34, 130, 141, 146–47,
        158, 166, 170, 185, 192, 247

    technological innovation by, 4, 110, 133,
        146–47
Aquaduct bike, 204–5, 282
Aquariass, 281–82
architecture, 9, 187–96, 201, 223–24, 234–35,
    280
Architecture for Humanity (AFH), 12–13,
    183–85, 187–90, 193–96, 208, 234, 284
Arent, Ben, 283
Arizona State University, 11, 72, 119–20, 164
art:
    design vs., 30–31, 269
    as part of design, 121–22, 196
arthritis, 27, 113–14
Arts and Crafts Movement, 5, 199–200
AskNature.org, 230
Aspen Design Challenge, 87–88
attention:
    directed, 75–78, 89, 92–93, 109–10
attraction, 134, *135*
    see also appeal
Au, Irene, 145–46, 177
Automotive X Prize, 285
awareness campaigns, 1, 2, 13, 150–52, 282
    see also posters
Axe, 158
Azizi, S. Ausim, 265

backpacks, 199
back pain, 267
Ban, Shigeru, 223–24
banking, 15, 126–30, 133–34, 140, 141–43,
    144–45, 147, 174–75

Bar, Moshe, 34–35
Bartels, Kathleen, 90
Bauhaus, 199–200
Baumgart, Ted, 250–51
beauty, 43–44, 228, 279
    functionality and, 35, 122, 258
Beckert, Oliver, 281–82
Before & After, 65
beginner's mind, 24–28, 52–55, 101
beginning anywhere, *14*, 16, 269–72,
    273, 287
Behar, Yves, *29*, 38, 118, 149–50, 152,
    161, 162, 233, 270
    XO laptop by, 10, 12, 16, 35, 45, 55–61,
        66–69, 216–21, 284
behavioral appeal, *33, 33*, 34, 35–36
Bennett, Paul, 58–59, 112
Benyus, Janine, 229–30, 277
bicycles, 2, 92, 204–5, 227, 280,
    282, 283–84
Bielenberg, John, 196
Bierut, Michael, *29*, 30, 37, 39, 62,
    77, 104, 258
Biloxi Model Home program, 193–95
biofuel, 213
biology, 229–30, 246, 277
Biomimicry: Innovation Inspired by Nature
    (Benyus), 229–30, 277
Birkenstock, 57
Blackwell, Marlon, 193–95
Blair, Maria, 186, 207, 209
BMW, 149
Bogusky, Alex, 13, 150–53, 284
book design, 9, 50–51, 89, 137, 202
bookstores, 51–52, 138–39
boredom, 264–65, *265*
Bowling Alone (Putnam), 253
BP (British Petroleum), 80
brainstorming, 272, *273*
Brand, Stewart, 17
Brandenburg, Kathleen, 106, 124, 254
branding, 7, 15, 25, 78, 80–83, 84, 126–54,
    161–62, 202, 204
Brandis, Jock, 211–14, 221, 233, 234, 285
Breuer, Marcel, 200
bridges, 209
Brooks, David, 141, 289
Brown, Tim, 39, 40, 74–75, 115, 125, 140, 274
Bruce, Gordon, 118
building to think, 74–75
bureaucracies, 68, 86, 90, 203
Burger King, 153
Bush, George W., 202–3, 204

business sector, 6–8, 13, 14–15, *14,* 23–24, 32,
    94–95, 99–180, *102*
    experience design by, 126–54
    internal culture of, 155–80
    research for, 99–125
butt brush effect, 117, 140
Buxton, Bill, 73–74, 180
Byrne, David, 280

CAD monkeys, 187, 196
Cage, John, 269
Campbell, Joseph, 143
Cannon, Charles, 62
cardboard, 223–24
careers, 187, 196, 198
carpets, 226
cars, 45, 100–101, 109–10, 123, 200, 202,
    227–28, 231–34, 244, 245, 251, 256, 283,
    285
    Mini Cooper, 1–2, 12, 148–50, 152–53, 231,
        233, 284
Carter Library, 212
Casey, Valerie, 36, 163, 179, 226, 235
celebrities, design by, 279–81
cell phones, 77, 117
Census Bureau, 86–87
chairs, 6, 31, 37, 183, 184, 190, 280
Challenger, 87
charts, 262, *263*
cheap, inexpensive vs., 56, 60
Cheetah (prosthesis), 48–49, 62, *63*
Chicago Cubs, 135–36
China, 117, 185
Chipchase, Jan, 117
chunking, 77, 93
circular processes, 269–72, *273*
Citibank, 25
cities, improvement of, 10, 15, 37, 43–44,
    89, 284
Civic Ventures, 214
ClearRX, 27, 83–86
Clinton Global Initiative, 205
clocks, 276–77
Clocky, 276–77
Clow, Lee, 158–59, 166–71, *169*
coats, 205–6, 284
Coca-Cola, 10, 15, 94, 153, 155–57, 161–63,
    164–66, 290
coffee, 143–44, 277, 278
Cohn, Neil, 79
collaboration, 6, 38–39, 44, 51, 74, 172, 175–76,
    193, 203, 213, 218, 230, 234–36, 274–75,
    290–91

Collins, Brian, 9, 14, *29, 72–73*, 78, 80–83, 147,
      153, 201, 204, 243–44, 270, 282, 285, 289
communication, 44, 70–95, 144–45, 203, 291
    between customers, 105, *133*, 147–48,
          156–59, 160, 170, 179–80
community, 44, 68, 81, 127, *133*, 142, 145, 191,
      208, 244, 247, 251, 252–58, 265, 283
compartmentalizing, design vs., 155–80
Compelling Experience Framework,
      134–37, *135*
complexity, 248–50
computers, 1, 2, 10, 12, 16, *35*, 45, 55–61,
      66–69, 108, 110, 146–47, 216–21
concrete, *213*, 214
Condé Nast, 179
connections, serendipitous, 61–66
consequences, *14,* 15, 183–210
    unexpected, 207–10, 241, 243
constraints, embracing of, 6, *14,* 15–16, 55–61,
      187–89, 197, 211–36, 243, 291
consumerism, 5–6, *183*, 186–87, 196, 203
consumers, scrutiny by, 94, 105, 156–59, 160,
      179–80
Continuum (design firm), 57, 58, 59, 116,
      174, 185
control panels, *33*, 108
Cooke, Marty, 177
cookware, 113
Corá, Gabriela, 265
Corning, 113
Corporate Social Responsibility (CSR)
      departments, 165
corporations:
    culture of, 155–80
    history of, 161–62, 167
    see also business sector
cosmetics, 115
Cottam, Hilary, 235–36, 256–57, 282–83
Cradle to Cradle (McDonough), 225–26,
      230, 231
craft, 200
Creative Artists Agency, 184
Creature, 129
criminal justice, 11
Crispin Porter + Bogusky (CP+B), 152–53
crossword puzzles, 62, *63*
Crow, Michael, 120
Csíkszentmihályi, Mihály, *143*, 249, 264
culture, 139–41, 152
    of corporations, 155–80
Curtis, Jamie Lee, 281
Cusato, Marianne, 190–93, 194, 195,
      244–45, 251, 253

customer loyalty, 135, 141, 147
customer service, experience design vs., 7, 130

Dair, Tom, 132
Daley, Richard, 94
data, information vs., 77, 85
Davis, Ray, 126–30, 144–45, 174–75
Day-Lewis, Daniel, 280
de Bono, Edward, 63–64
democracy-of-design, 5, 44, 249
Dempsey, Mike, 37
dentistry, 28
Denver Biennial of the Americas, 290–91
design:
    broad application of, 4–6, 9–11, 36–38, 90–95,
          185
    definitions of, 2, 3, 26–32, *29,* 37, 157, 241–42
    history of, 5–6, 31–32, 199–200
    profitability of, 106–7
    resistance to, 243
    schools for, *see* schools, for design
design activism, 183–210
Design and the Elastic Mind (museum show), 94
Design Council, 107
designed experiences, 24, 37, 126–54, 161, 172,
      245
Designers Accord, *163*, 226, 235
"Designer's Road to Hell, The"
      (Glaser), 201–2
Design for the Other 90% (museum show), 94,
      221–22
Designism 3.0, 203–7, 285
design movements, 5–6, 199–200
Design Observer blog, 208
design porn, 248
design thinking, 13–16, *14*
    for business, 14–15, 24, 32, 94–95, 99–180, *102*
    for personal sector, 13, 16, 24, 237–66
    for social sector, 15–16, 24, 29–94, 183–236,
          285
    universal, 13–14, 21–95, 271–73
detergent, 117, 171–73
diabetes, 239, 277, 278–79
diapers, 281
dinosaur babies, 58–59, 66–69
directing attention, 75–78, 89, 92–93, 109–10
disasters, averting of, 79–88, 209
disengagement, 65–66, 253, 288–89
distributed possibility, 202–7, 243
distribution, 66–69, 206–7, 213–14
Doblin, 103–4, 134–35
dogs, 2, 15, 166–71, 174, 193, 195, 268
Dove soap, 80, 158

Downsview Park (Toronto), 51
Dreyfuss, Henry, 108
d.school, 6, 103, 110, 178, 215, 269, 271–75
Duffy, Joe, *29*
Dun, Niki, 283
durability, 60, 224–26
Dyson, James, 114

Eames, Charles, *29,* 37, 107, 118, 163, 200–201
Eames, Ray, 200
earned media, 171, 286
Eastern philosophy, 260
eco-homes, 245, 250–51
economic crises, 2, 3, 5, 7, 15, 23, 87, 99, 103,
    127, 177, 183, 187, 209, 214, 239, 291
Edible Estates, 252
Edison, Thomas, 288
education, 15, 56, 66–67, 70–72, 91–92, 135,
    236, 247, 257–63, 284, 288, 290
    see also schools, for design
efficiency, 99, 127, 196–97, 217, 221, 223, 241
egocide, 200, 236
emergence, *14,* 16, 239–66, 288, 289
emerging markets, 221–23
Emotional Design (Norman), 35
empathic research, 7, 9, 13, 99–125, *102,* 127,
    131, 167, 172, 178, 186, 193, 196–99, 254,
    268–69, 271, 272–74, *273*
empathy, *3,* 7, 12, 13, 22, 40, 285–86
empowerment, 239–91
Encyclopedia of Life, 230
energy, 217, 227–28, 231, 233, 244, 251
    sustainable, 6, 58, 59, 92, 194, 204–5, 212,
        213, 222, 245, 250–51, 285
engagement, 134, *135,* 253
engineers, 12, 172
environmentalism, 5, 6, 15, 43, 44, 51, 79–81, 92,
    94, 155–64, 179–80, 187, 201, 209, 215,
    224–28, 244, 250–52
ephemeralization, 216–18
ergonomics, 27, 107–8, 114
ESI Design, 187, 254
ethics, 83, 124, 201–2
Euro RSCG, 179
experiences:
    designing of, 117, 126–30, 134–35, 140–45
    mapping of, 134–35
    objects vs., 24, 37, 126–54, 161, 172, 245
experimentation, 50–55, 74–75, 172, 176–79
expertise, as liability, 24–28, 52–55, 259–61
exploratory sketching, 74
extreme users, 114–15
eyeglasses, 221

Facebook, 4, 133, 147, 247, 283
Fahnstrom, Dale, 75
failure, 5, 50, 67–68, 209
familiarity, 76, 78, 141–42
Farber, Sam, 27
farming, 221, 222
fashion design, 9, 28, 31, 205–6
featuritis, 7, 57, 100, 104, 131, 146, 240
feedback, 6, 54, 66–69, 75, 100–101, 176, 220,
    274–75
FEMA (Federal Emergency Management
    Agency), 190–91, 195
Fibonacci spiral, 246, *246,* 258
"15 Below" coat, 205–6, 284
findability, 76–78
firefighters, 273
Fisher, Martin, 222
Fisher, Thomas, 209, 284
Five Whys, 25, *26,* 274
Florida Neuroscience Center, 265
flow states, 135, 143, 264–66, *265,* 288–89
focus groups, 100–101, 111
fog of war, 53, 119
food, 240, 252
Ford, Henry, 100–101
form, 229–30
Formosa, Dan, 113, 148
fracture critical, 209, 244, 260
framing, 21–28, 272, *273,* 274, 279
Frandsen, Mikkel Vestergaard, 282
Fraser, Heather, 100, 102, 177
Friedman, Thomas, 205
friendship, 254–58
    see also collaboration; community; Internet
Frisbee, 282
Full Belly Project, 214
Fuller, Buckminster, 5, 41, 122, 200, 217, 234,
    255, 291
Fulton Suri, Jane, 13, 108–11, 123, 273
functionality, 27, 30, 33, 34, 57–58, 107–8,
    131–32, 145–46, 217
    beauty and, *35,* 122, 258
furniture, 198, 280
    see also chairs
fuseproject, 57

Gadget Nation (Greenberg), 279
gadgets, golden age for, 281–82
Galaxy Alliance (font), 81–82
games, 253, 254–55
gardens, 252
Gatorade, 106
Gehry, Frank, 9, 51, 161–62

General Electric, 229
George Brown College, 90–92
geothermal heat, 245
"get-to" vs. "getaway," 252–54
Glaser, Milton, 28, 29, 38, 47, 74, 78, 85, 175,
    201–2, 204, 242
glimmer moments, 2, 13, 46, 62, 65, 198
glimmer principles, 13–16, 14, 271–73, 273
glo Pillow, 277
"going deep," 14, 15, 99–125, 102, 130, 167,
    185, 193, 198–99, 209, 211, 254, 256, 260,
    273, 277, 287
Goldenberg, Seth, 290
golf courses, 279
good, forgetting about, 10, 64–65, 288
Google, 145–46, 147, 177–78, 247, 280–81
Gore, Al, 79–82, 187
governments, 2, 3, 10, 86–87, 94, 108, 184, 190,
    203
graphic design, 9, 28, 42, 201
grassroots change, 68, 202–7
Graves, Michael, 3, 5, 40, 140
Greenberg, Bob, 133
Greenberg, Steve, 279
Greene, Todd, 277, 278, 279
Grefé, Richard, 86–87
Gropius, Walter, 200
growth, 245–47, 246, 258–61, 288
Guatemala, 3, 10, 94, 119
Gulf Coast, 190–95

habits, 8, 50, 149
Hackett, Jim, 178
Haeg, Fritz, 252, 257–58
hand tools, Stone Age, 114, 214
happiness, 16, 31, 124, 249, 251, 255, 261,
    264–66
Harvard University, 34, 253
Hasher, Lynn, 259
Haven, Jim, 129
Hayward, Lani, 141
HeadBlade, 277, 278, 279
health care, 79, 83–86, 184, 188–89, 239–40,
    283, 285
    see also prosthetics
Heatherwick, Thomas, 38
Helvetica, 82, 151
Helvetica Neue Bold Condensed, 93
Herman Miller, 57, 229
heuristic bias, 8, 63–66, 149
Hewlett-Packard, 124, 289
Hildebrand, Gert Volker, 149
Hippo Roller, 209–10

Hockney, David, 30
homelessness, 8, 12–13, 24, 198–99, 205–6, 284
homes, 245, 247, 248–52
    see also shelter
hospitals, 37, 112, 188–89

IA Collaborative, 28, 29, 106, 254
iBOT, 10, 21–23, 27, 92, 285
IDEO, 6, 55, 173, 177, 185, 186, 198, 222, 254,
    271, 283
    designers from, 36, 39, 58–59, 74, 132, 140,
        163, 179, 204, 226
    Five Whys of, 25, 26, 274
    research methods of, 13, 110–12, 115, 116,
        118, 120, 123, 125, 273
ID Two, 110
ignorance, 207–8
    benefits of, 24–28, 52–55, 259–61
Illinois Institute of Technology (IIT), 74–75, 103,
    104, 171, 172, 173, 175
iMacs, 166
immersive experiences, 135, 143, 264–66
improving, inventing vs., 36–38
Incomplete Manifesto for Growth, An (Mau), 10,
    13, 54, 65–66, 101, 128, 269–70, 287–89
Inconvenient Truth, An (Gore), 79–80
incorporation, 159–64
INDEX: AWARD, 285
Indigo (bookstore), 51–52, 138–39, 290
Indonesia, 184
inept users, 114–15
Information Anxiety (Wurman), 261, 262
information processing, 75–80, 84–85,
    93, 146, 259
infrastructure, 209, 214, 222
"In Good We Trust" (Denver Biennial), 290–91
InnoCentive, 205, 285
innovation, 45–69, 73–74, 103–4, 186, 235
innovation gap, 7, 99–125, 105
instinct, research vs., 101, 107, 118–20, 145–46
Institute without Boundaries, 91–92
integrated marketing, 12, 114, 133, 147–50,
    152–54
integrative thinking, 61–66
Intel, 58, 61, 67
interconnectedness, 160, 180, 225, 255, 265
interdisciplinarity, 39–40, 41, 50, 62, 175–76,
    258–61
interior design, 16, 28, 31
International Contemporary Furniture Fair, 183
Internet, 77, 94, 145–46, 147–48, 152, 158, 203,
    205, 206, 214, 230, 285, 290
    social networking on, 4, 133, 234–35, 283, 284

invention, 36–38, 61–66, 275–76, 288
iPhone Developer's Cookbook, The (Sadun), 77
iPhones, 34, 77, 147, 150, 166
iPods, 34, 35, 105–6, 132–33, 146, 147, 166, 177, 192, 203, 230, 231, 253, 281
Iraq War, 46
irrigation, 221, 222
Isley, Alexander, 9, 28, *29*
iTunes, 247
Ive, Jonathan, 101

Japan, 172, 224
Java Jacket, 277, 278
Jawbone headphones, 57
Jobs, Steve, *29*, 101, 146–47, 166
Judd, Donald, 30
Jump Associates, 120–23
"jump fences," 14, *14*, 45–69, 127–28, 175–76, 212, 233–34, 259, 268, 271, 287
Jung, Carl, 143
Jung-Beeman, Mark, 65

Kalman, Tibor, 64
Kamen, Dean, 10, 12, 21–23, 27, 38, 45, 46–47, 55, 56, 61, 92, 205, 216, 236, 282, 285
Katrina, Hurricane, 5, 71–72, 187, 190–96
Katrina Cottage, 190–93, 195, 244
Katrina Furniture Project, 190
Keeley, Larry, 103–4, 135–36
Kelley, David, 110–11, 132, 271
Kelley, Tom, 111
Kembel, George, 178–79, 215–16, 271–75, 282
KickStart, 222
kitchens, 196–97, 248
knowledge, curse of, 24–28, 52–55, 259–61
Koolhaas, Rem, 9, 51
Kosovo, 187, 188
Kotchka, Claudia, 172–74

labor practices, 158, 160, 168–69, 171, 174, 202
Lacayo, Richard, 93, 201
Lafley, A. G., 171–74
laptops, low-cost, 1, 2, 10, 12, 16, 35, 45, 55–61, 66–69, 216–21, 284
lateral thinking, 63–66, 234
lawns, 252
Le Corbusier, 200, 201
Lehman Brothers, 209
Leinen, Eleanor, 277, 278–79
life planning, 28, 239–66, 279
LifeStraw, 92, 157, 221, 282
Linux, 67
Loewy, Raymond, 37, 162

logos, 25, 78, 80–83, 84, 136, 202, 204
Lois, George, 25–26, *29*, 62
"lost in the woods," 52–54, 92, 101, 260, 289
Lupton, Ellen, *29*, 30

Ma, Yo-Yo, 206
McCallion, Steve, 127, 141–43, 147
McDonough, William, 225–26
McGonigal, Jane, 255
Macintosh computers, 4, *173*
McLaren F1 supercar, 231, 232, 233
Maeda, John, 39, 78, 199, 260, 261
Mali, 211–13
mapping, 35, 58, 108
Mars, Inc., 166
Martin, Kerri, 148–50, 152
Martin, Roger, 7–8, 47–48
Massachusetts Institute of Technology (MIT), 56, 61
Massive Change exhibition, 4, 10, 12, 17, 36–38, 55, 89–95, 120, 160–61, 185–86, 201, 204, 221, 229, 234, 245, 290
mass-to-information ratio, 83–85
Masterfoods, 166
mastery, *33*, 34, 35–36, 143
materials, 188, 190, 212, 219, 223, 225, *225*, 232, 234, 235, 251
Mathieu, Marc, 155–56, 161
Mau, Bruce, 36, 47, 50–55, 68, 70–72, *73*, 75, 88, 203, 207, 239–43, 247, 249, 250, 261, 266, 270, 272, 289–91
    background of, 9–12, 40–42
    on business, 23–24, 119–20, 121–22, 137–39, 153–54, 155–66, 176
    design defined by, *3*, 5, *29*, 30, 31–32, 241–42
    environmentalism of, 155–56, 224–28, 229, 245, 265, 282
    laws of, 10, 13, 54, 65–66, 101, 128, 269–70, 287–89
    Massive Change exhibition of, 4, 10, 12, 17, 36–38, 55, 89–95, 120, 160–61, 185–86, 201, 204, 221, 229, 234, 245, 290
    on questioning, 10, 22, 23–24, 27
    speeches by, 43–44, 121–22, 227–28
maximalism, 248–50
measuring cups, 131–32
media, earned, 171, 286
Medicare, 86–87
medicine bottles, 1, 2, 27, 83–86, 198
Merholz, Peter, 146
metaphors, *14*, 15, 126–54, 287
methodology, 271–75, *273*, 287–89
Method soap, 99

Michigan, University of, 235, 259

Microsoft, 58, 61, 67, 73, 147, 153, 180

military technology, 46–47, 107–8

Mini Cooper, 1–2, 12, 148–50, 152–53, 231, 233, 284

Minnesota, University of, 198–99, 209, 284

MIT (Massachusetts Institute of Technology), 56, 61

mobile health clinics, 188–89, 283

modeling, 3, 14, 28, 37, 45, 52, 66, 70–95, *102*, 155–56, 172, 183, 189, 198, 203, 220, 223, 254, 262–63, *263*, 271, 291
    see also prototyping; sketching

modernism, 5, 6, 198–99, 201, 226, 248

Moggridge, Bill, 110

Mok, Clement, 25, *29*, 72

MoMA design store, 277

Monahan, Tom, 65

mops, 116, 172–73

Morris, William, 5, 199–200

mortgage contracts, 87, 248

motorcycles, 109–10

Motorola, 273

MP3 players, *see* iPods

MTV, 11, 94, 164, 290

multiplayer games, 255

Murray, Gordon, 45, 231–34, 283

Museum of Contemporary Art (Chicago), 161

Museum of Modern Art, 94, 215

museums, 51, 70–72, 89–95, 119, 161–62, 249, 254

Mycoskie, Blake, 286

Mykolyn, Steve, 205–6, 284

Nadal, Miles, 239

Nanda, Gauri, 276–77

National Design Award, 286

National Design Triennial, 93–94

natural disasters, 12–13, 16, 183–85, 190–95, 224

needs, human, 6–8, 12, 24, 28, 36, 45, 99–125, 183–210, 214, 279
    wants vs., 105, 124–25

Negroponte, Nicholas, 56–61, 66–67

Nelson, George, 80

neuroscience, 32–36, 57, 65, 74, 75–76, 78, 160, 259, 264–65

New Orleans, La., 70–72

New York, N.Y., 47, 77, 78, 280

Nike, 4, 99, 105–6, 132–33, 158, 229

9/11, 187, 189, 202

Nissan, 158

Nokia, 94, 99, 117, *133*, 138

Noonan, Mark, 267–69, 274, 276

Norman, Donald, 32–33, *33*, 34, 118, 140

North Star principles, 144–45, 166–71

Northwestern University, 32, 65

novelty, 78, 142

Noyes, Eliot, *29*, 107, 118

Nussbaum, Bruce, 67, 103, 283

nuts, shelling of, 211–13, *213*, 221, 285

objects, 157, 160
    experiences vs., 24, 37, 126–54, 161, 172, 245

obsolescence, planned, 6, 186–87, 201, 224–26, 250

Ogilvy & Mather, 80

Oil of Olay, 173

Olson, Elliott, 197

180-degree thinking, 65

One Laptop Per Child (OLPC), 10, 12, 16, 35, 55–61, 66–69, 216–21, 284

Ontario College of Art & Design, 41, 245

Open Architecture Network (OAN), 234–35

openness, 155–80, 254, 290–91

open-source model, 67, 189, 203, 213

optimism, 2, 16–17, 47–48, 81, 93, 99, 177, 200, 201, 270, 291

Oral-B, 116

out-of-box experiences (OOBE), 132

OXO Good Grips, 27, 113–14, 123–24, 131–32, 175–76

packaging, 1, 2, 27, 32, 83–86, 100, 161, 162, 165, 202

Page, Scott, 235

Panama, 89–90, 119

Papanek, Victor, *29*, 242

parks, 44, 51

Participle, 235, 256

patents, 213, 274, 280–81

Patnaik, Dev, 120, 123

patterning, 8, 62–66, 76, 78, 291

Patton, Phil, 5

Peace Corps, 211, 214

peanuts, 211–13

Pedigree, 15, 166–71, 174

peelers, 27, 113–14

Peer Insight, 106–7

Pentagram, 24, 30, 37, 39, 41–42, 62

People's Design Award, 192

personal applications, 8, 13, *14*, 16, 24, 237–66

Philips, 112

Phillips, Van, 48–49, 62, *63*

pictograms, 79, 85, *85*, 262

Pilloton, Emily, 184, 209–10, 282

pillows, 277
pirates, 136–37
pitfalls, 117–18, 207–10
Pitt, Brad, 280
pizza cutters, 115
play, 254–55, 274
politics, 42, 50, 202
Pommern, Tanzania, 197–98
population growth, 187, 209, 244, 255
Porchdog house, 193–95
porches, 191, 194
Portland, Oregon, 284
posters, 50, 201, 270
    see also awareness campaigns
Postrel, Virginia, 31, 32
poverty, 70–72, 92, 183–210, 215
Prahalad, C. K., 222–23
problem solving skills, 45–69, 255, 259
Procter & Gamble (P&G), 15, 99, 116–17, 158,
    171–74, 229
products, lifecycle of, 132, 224–26
Project H, 184, 209–10, 282
Project M, 196
prosthetics, 10, 45, 46–47, 48–49, 221
pro-thinking policies, 178–79
prototyping, 72, 74–75, 85–86, 113–14, 172, 178,
    209, 212–13, *213*, 219–20, 235, 270, 271,
    272, *273*, 274, 276–77, 279
    see also modeling; sketching
Prygrocki, Greg, 75
psychology, 32–36, 57, 107–11, 117, 134, 146,
    150–51, 249, 264–65
Public Good, 50
public housing, 201, 208
public transportation, 89, 228, 283
Putnam, Robert, 253
puzzles, 62, *63*, 78–79, 233, 262

Q Drum, 221
questioning, 8, 10, 14, *14,* 21–28, 45, 130, 175,
    185, 203, 229, 240, 242, 271, 272, 274, 287
    examples of, 127, 150, 231–32, 244, 251–52,
        256, 261
    Mau on, 10, 22, 23–24, 27

Rabe, Cynthia Barton, 54
radios, handheld, 144–45, 273
rainwater collection, 197, 282
Rashid, Karim, 5, 6, 196
razors, 277, 278, 279
recombination, 4, 45–50, 61–66, 90, 175–76,
    185, 212, 219–20, 234, 259, 267, 281–83,
    288

Recor, Bret, 217–18, 220
recycling, 6, 45, 132, 162, 163, 165–66, 225–26,
    230, 231, 232, 251
reflective appeal, *33, 33, 34*, 140–41, 151, 159
research, 45, 134, 208, 277
    empathetic, 7, 9, 13, 99–125, *102*, 127, 167,
        172, 178, 186, 193, 196–99, 254,
        268–69, 271, 272–74, *273*
    as secondary, 52–55, 58, 118–20, *119*, 145–46,
        270, 271
R/GA (firm), *133*
Rhode Island School of Design (RISD), 39, 62,
    78, 196–99, 260
Ritz-Carlton, 144–45
Rockefeller Foundation,, 186, 207, 209
Rockwell, David, 249
Rotman School of Management, 7, 47, 100, 102,
    103, *173*, 177, 269
Royal Society of Art, 277
Rubbermaid, 115–16
runners, 105–6, 132–33
Rwanda, 224

Sagmeister, Stefan, *29*, 30, 63–64, 65, 289
Salvation Army, 194–95, 206
Scher, Paula, 24–25, *29*, 30, 62, 269
Schlossberg, Edwin, 187, 202–3, 254–55
Schmidt, Eric, 177–78
School of the Art Institute of Chicago,
    88–89
School of Visual Arts, 72, 83
schools, for design, 4, 6, *39*, 83, 88–89, 90–92,
    173, 196–99, 262, 271–75
    see also education
science, 121–22, 229–30, 246, 277
Seattle Public Library, 9, 51
Segway, 12, 48, 55
Seligman, Martin, 264
Senate, U.S., 192
Shaw Industries, 226, 290
Shedroff, Nathan, 270
shelter, 8, 10, 12–13, 184, 187–88, 189–96, 224,
    235, 284
Shephardson Stern + Kaminsky, 177
Shimano, 283
shipping, 159, 163, 165
shoes, 4, 99, 105–6, 132, 277, 278–79, 286
signals, 76, 139
Silicon Valley, 57, 58, 60, 61, 67, 112
Simon, Herbert, 242
Simons, Daniel, 76
simplifying, 57–58, 61, 73, 77–78, 81, 124, 145,
    146, 191, 211–36, 248–50

Sinclair, Cameron, 12–13, 16, 183–85, 187–90,
    193–96, 201, 208, 234–35, 284
Six Sigma, 103
sketching, 73–75, 189, 231, 269, 270
    see also modeling; prototyping
skids, steering into, 177–78
Skype, 214, 241
sleep, 65–66, 240, 276–77
Smart cars, 233
Smart Design (firm), 32, 113–14, 132, 148
smartphones, 77, 253
smart recombinations, 4, 45–50, 61–66, 90,
    175–76, 185, 212, 219–20, 234, 259, 267,
    281–83, 288
Smithsonian Institution, 192
smoking cessation, 1, 2, 13, 150–52
snow shoveling, 267–69
social networking, 4, 133, 234–35, 283, 284
social responsibility:
    of corporations, 155–80
    of designers, 29, 183–210
    of individuals, 243–45, 265
    see also environmentalism
social scientists, 107–11, 273
social sector, 8, 13, 14, 15–16, 24, 79–94, 183–
    236, 285
Social Security, 86–87
Sohn, Michelle, 132, 176
solar power, 59, 250–51, 285
Sorensen, Jay, 277, 278
Southwark, London, 256–57
speculation, 52–55, 119
sport-utility vehicles (SUVs), 148, 202
Spradlin, Dwayne, 205
Sri Lanka, 189
Stanford University, 6, 103, 110, 178, 215, 269,
    271–75
Starbucks, 34, 140, 143–44, 145
starchitects, 204
Starck, Phillipe, 3, 5, 6
Steelcase, 178
Stewart, Jon, 206
Stone Age hand tools, 114, 214
store brands, 167, 171
store design, 117, 126–30, 134–35, 140–45
Stowell, Davin, 32, 113–14
Strategic Innovation Lab, 245
strategic planning, 102, 102, 155–80
structural reorganization, 102, 155–80
style, 51
    design vs., 3, 30, 31–32, 73, 157
Substance of Style, The (Postrel), 31
suburbs, 244–45, 251–52, 253

subway map, of New York, 47, 77
Sudbury, Ont., 40–41, 43–44
Sugden, Kevin, 261
surveys, 100–101
Sussman, Laura, 197–98
sustainability, 36, 44, 132, 155–64, 208, 215,
    224–28, 230, 231–34, 248–52, 291
Swash, 117, 173
Swiffer, 172–73
synthesizing, 61–66
systems design, 242, 255

T.25 (car), 231–34
tables, 198, 199
Tanzania, 197–98
Target, 5, 15, 86, 130, 140, 141, 171
Taxi (design firm), 205
TBWA\Chiat\Day, 166–71
technological advancement, 46, 61, 94, 107–8,
    132–33, 203, 218, 234, 281
    innovation vs., 104–5, 146
TED conferences, 187, 261–62, 291
temporality, 132–33, 134–35
Thackara, John, 61, 62, 203, 208, 219
Thatcher, Margaret, 42
theatricality, 130, 137, 152–53, 155
"Think Different" campaign, 166
thinkering, 74–75, 173, 178, 274
Three Gears of Design, 102, 102, 120
Three Mile Island, 33
Tide, 117, 172
Timberlake, Justin, 279
tinkering, 74–75
toilets, 281–82
Tongue Sucker, 285
Total Quality Management (TQM), 103
touch points, 133–34, 149–50, 152–53
touch screens, 219
TQM (Total Quality Management), 103
trailers, 190–91, 195
transformational metaphors, 127–30, 140
transformation design, 4–5, 9–11, 36–38, 39–40,
    166–71, 174–80
transparency:
    of organizations, 155–80
    of products, 213, 220, 232
Trollbäck, Jakob, 203
"Truth" campaign, 13, 150–52, 284
T-shaped people, 38–40, 62, 259–61
tsunamis, 187, 190
Tufte, Edward, 87
Tulane University, 70–72
Turner, David, 161

Turner Duckworth, 161, 162
12 Steps on the Designer's Road to Hell, 201–2
Tyler, Richard, 195
typography, 80–83, 84, 93, 151
"tyranny of the visual," 31–32

Uganda, 117
Umpqua, 15, 126–30, 133–34, 140, 141–43, 144–45, 147, 174–75
Underhill, Paco, 117
Understanding Healthcare (Wurman), 79, *263*
unfocus groups, 112
Unilever, 118, 158
uniqueness, 60–61
United Nations, 185
universal nut sheller, 211–13, *213,* 221, 285
universal principles, 13–14, *14,* 21–95
UPS, 164
upward spirals, 47–48
urban personal transport, 231–34
user-centered design, 99–125, *102,* 131–34, 161, 186
utopianism, 5, 198–99, 201

vacuum cleaners, 109, 114
value, comparative, 222
Van Alstyne, Greg, 92, 104, 245–46, 247, 248, 258
Vancouver Art Gallery, 90, 92–93
Van Halen, Eddie, 281
vending machines, 111, 129, 161
Venice Biennale, 31, 37
Vignelli, Massimo, *29,* 47, 77, 186–87
Virgin Airways, 130
visceral appeal, 32, *33, 33,* 34–35
visual processing, 31–32, 75–77, 78, 139
voluntourism, 253
Vossoughi, Sohrab, 174
voting ballots, 86–87
voyeurism, 117–18

walkers, 115–16
walkie-talkies, 144–45, *273*
Wal-Mart, 158, 171
Walt Disney Concert Hall, 9, 51
Walton, Ian, 277
Ware, Colin, 75–76, 78
Warhol, Andy, 162
warning icons, 85, *85,* 215

waste, 99, 186–87, 191, 217, 225–26, *225,* 231, 250, 282
   as anti-advertising, 155–57, 158
watches, 149–50
water:
   conservation of, 87–88, 164, 197, 230, 244, 252, 282
   purification of, 2, 10, 92, 156, 184, 186, 204–5, 207, 211, 221, 282, 285
   transportation of, 2, 204–5, 209–10, 221, 282
   see also irrigation
wayfinding, 93, 141, 245
   see also attention, directed
Web design, 9, 28, 77, 133
wheelchairs, 1, 2, 10, 21–23, 27, 92, 285
Whitney, Patrick, 103, 104–5, 171, 172, 173, 175, 178
Whole Foods,, 140
"why," importance of, 24, 25–27, *26,* 274
wicked problems, 207–8
Williams, Bisi, 89
Windows, 147
wind turbines, 6
Wired to Care (Patnaik), 123
Wolff, Michael, 204
Wong, Liliane, 197, 198
Wong, Robert, 34
work-arounds, 115–16
"working the metaphor," *14,* 15, 126–54, 287
work spaces, *173,* 175–76, 289
World Game, 255, 291
World War II, 107–8
Worldwatch Institute, 253
Wovel, 267–69, 274, 276
Wrigley Field, 135–36
Wurman, Richard Saul, 79, 261–64, *263,* 265, 288

XO laptops, 10, 12, 16, 35, 45, 55–61, 66–68
XO-2 laptops, 69, 216–21, 284

"Yes is more," 227–28
Young, Kevin, 58
YouTube, 4, 11, 203, 204

Zaccai, Gianfranco, 174
Zero Gravity Thinkers, 50–55
Ziba Design, 127–28, 141–43, 147, 174–75
Zone books, 9, 50–51, 89, 137, 159–60

Illustration Credits

p. 26. © IDEO Method Cards

p. 29 © IA Collaborative

p. 49 © Van Phillips

p. 59 © Yves Behar

p. 70 © Bruce Mau

p. 71 © Bruce Mau

p. 79 Designed and published by Richard Saul Wurman

p. 82 © Brian Collins

p. 85 © Target Corporation. Designers: Milton Glaser and Deborah Adler.

p. 88 © Brian Collins

p. 90 © Bruce Mau

p. 102 © Heather Fraser, Roger Martin Designworks™, Rotman School of Management

p. 114 © SmartDesign

p. 116 Oral B toothbrush image courtesy of IDEO

p. 119 © Bruce Mau

p. 128 © Ziba Design

p. 132 © SmartDesign

p. 135 © Larrry Keeley, Doblin Inc.

p. 136 © Brian Collins

p. 138 © Bruce Mau

p. 151 © Alex Bogusky/Crispin Porter + Bogusky

p. 169 © Lee Clow

p. 192 © Marianne Cusato

p. 194 © Marlon Blackwell

p. 199 © Sohyun (Katie) Kim and Haemin Hong

p. 202 © Milton Glaser

p. 206 © TAXI

p. 218 © Yves Behar

p. 223 KickStart image courtesy of IDEO

p. 232 © 2009 Gordon Murray Design Limited

p. 263 Designed and published by Richard Saul Wurman

p. 265 © Mihály Csíkszentmihályi

p. 268 © Structured Solutions II, LLC

p. 276 © Gauri Nanda

p. 278 © Todd Greene

p. 280 Organic Pads, 2004, Pen on paper. By David Byrne, courtesy Pace/MacGill Gallery

p. 280 Bike rack by David Byrne